AF553658

AVIAN ENDOCRINOLOGY

ENCYCLOPAEDIA OF ENDOCRINOLOGY-IV

AVIAN ENDOCRINOLOGY

By

Manju Yadav

Lecturer

Department of Zoology

M.M.H. College

Ghaziabad (U.P.)

(India)

DISCOVERY PUBLISHING HOUSE PVT. LTD.

NEW DELHI-110 002

First Published-2008

ISBN 978-81-8356-274-4

© Author

Published by

DISCOVERY PUBLISHING HOUSE PVT. LTD.

4831/24, Ansari Road, Prahlad Street,
Darya Ganj, New Delhi-110002 (India)
Phone: 23279245 • Fax: 91-11-23253475
E-mail: dphbooks@rediffmail.com
dphtemp@indiatimes.com

Printed at:

Sachin Printers, Delhi

Preface

The present title **Avian Endocrinology** provides a definitive and comprehensive review of all aspects relating to hormones in different animals including wild and domestic species. It discusses the intimate physiology of the endocrine system itself and describes the role of hormones in the processes of nutrition, osmoregulation, colour change, calcium metabolism and reproduction. Subject matter has been presented in a readable way emphasizing the differences between species from a functional aspects. It not only contains a wealth of information concerning the endocrinology of the various species but also contains more than enough basic information concerning the biosynthesis, metabolism and mode of action of the hormones for the book to be read and understood without referring to a text book of basic endocrinology. This book has been written primarily for use as a text book by under-graduate, as well as graduate students.

Efforts have also been made to make the presentation lucid and accessible for beginning students, emphasizing major concepts while in corporating the latest experimental findings and theories throughout. The aim is to enthuse the readers with this active and exciting area of research and to lay a solid foundation on which further study of its various facts may be based.

Though, the author has taken special care to present a current account, yet he is fully aware about limitations and the readers may come across the mistakes of various types. For all types of mistakes author extends his due apology.

There can be no claim to originality except in the manner of treatment and much of the information has been obtained from the books and scientific journals available in the different libraries.

The author expresses his thanks to his friends and colleagues whose continue inspirations have initiated him to bring out this book.

The author expresses his gratitude to Mr. Wasan and staff of M/s Discovery Publishing House for their whole hearted co-operation in the publication of this book.

Author

Contents

1

Pituitary Gland

The aim of this chapter is to correlate the morphology of the pituitary cells with physiological analyses of pituitary function and chemical studies on the pituitary hormones insofar as they have been isolated. The past 15 years have seen great progress made into understanding these three facets of avian biology. In the field of morphology, the development of cytochemistry and the introduction of the electron microscope have permitted the easier identification of cell types and improved assessment of their functional state. During the same period there has been a rapid extension in our knowledge of avian endocrinology, and especially of avian neuroendocrinology. Chemical and immunochemical studies on the adenohypophyseal hormones have begun much more recently, but one can hope that in the near future immunocytochemistry will aid in the identification of the cells that produce various hormones.

At present, these cellular localizations can be established only on the basis of morphological data and must still be considered tentative. Such morphological correlations are nevertheless an important and necessary precedent for immunocytological studies. Moreover, unlike cytology, the chemistry of the avian pituitary hormones and the immunocytochemical identification of the cell types will inevitably be limited to a small number of species.

Anatomy and Embryology

Most of the basic information on the anatomy and embryology of the avian pituitary has been carefully and fully described by Wingstrand (1951), and we shall refer to his monograph on a number of occasions. Rather more attention will be paid, however, to recent studies concerned with cellular arrangement, pituitary vascularization, histogenesis, and the experimental analysis of pituitary development.

General Anatomy

The gross anatomy of the pituitary is remarkably uniform among birds. According to Wingstrand's nomenclature, the major part of the adenohypophysis is the pars distalis, which lies ventral or rostroventral to the neurohypophysis, and which in some species is divided externally into two parts by a transverse sulcus. The pars intermedia is not present as a separate entity. The pars tuberalis consists of three parts—the pars tuberalis proper, consisting of strings of cells covering part of the surface of the diencephalon; a portal zone, which consists of cells lodged between and along the portal vessels; and the pars tuberalis interna, which is intimately fused with the pars distalis. Depending on the species, these different parts may be developed unequally.

Microscopic Anatomy

Pars distalis

The glandular cells are arranged in cords or in acini and are limited by a basement membrane which often appears infolded under the electron microscope. Between the cells lies an important plexus of blood capillaries, the fine structure of which has been described for the White-crowned Sparrow (*Zonotrichia leucophrys gambelii*) by Mikami et al. (1970). They are of the typically fenestrated type as seen also in the rat pituitary and in other endocrine organs. The capillary wall consists of a single layer of endothelial cells containing many diaphragm-covered fenestrations. On their perivascular faces the endothelial cells are lined by a basement membrane. The majority of the glandular cells have direct contact with the pericapillary space, sometimes by means of narrow cytoplasmic extensions. The small chromophobic cells of the pituitary have no such relationship with the capillaries and lie in the center of the cell clusters. Under the light microscope, these appear to form aggregates of nuclei—the so-called *Kernhaufen* of German authors. Such figures are especially frequent in wild birds out of the breeding season.

Several histological features support the view that the pars distalis is divisible into a caudal lobe and a cephalic lobe. According to Rahn and Painter (1941) the most important difference between the lobes lies in the distribution of two types of acidophilic cells, the larger Al cells being restricted to the caudal lobe, while the A2 cells, which are smaller and contain many fine granules, are characteristic of the cephalic lobe. Wingstrand (1951) confirmed this subdivision in a great number of species, and we shall revert to this important question later. The cellular arrangement in the two lobes is also frequently different, with the caudal region being characterized by numerous pseudo-acini and the cephalic by compact clusters or longitudinal cords of cells. However, it is important to note that in no species studied is there a true separation of the two lobes either by a connective tissue wall or by a distinct alignment of the cells, the two lobes invariably being mixed in the median region. Moreover, the transverse sulcus of the pars distalis, whenever it exists, does not correspond with the histological subdivision.

A typical feature of the avian pars distalis is the presence in the adult of chromophobic and PAS (periodic acid-Schiff)-positive colloid droplets in the center of the acini. These formations are sometimes so numerous that the sections almost give the impression of being a thyroid gland. They have been described by several authors, and their significance has often been discussed. More recent studies with the electron microscope show that the colloid has nothing to do with any pituitary secretion and is a fibrous material arising from the cytoplasm of the apical zones of several gland cells.

Pars tuberalis

In the pars tuberalis proper, the flattened strings of cells are surrounded by a basement membrane. The preportal vessels course through these cell cords. They have the typical form of venous capillaries with a single layer of endothelial cells invested in a basement membrane. The junctions between these capillaries and the pars tuberalis cells are much less intimate than with the cells of the pars distalis. In the portal zone, the tuberal cells are arranged in long strings oriented with the blood vessels. The cytoplasm of the glandular cells is frequently almost chromophobic, although after Bodian impregnation, some cells show a few argyrophilic granules. To our knowledge no work has been published on these cells since that of Wingstrand (1951).

Vascularization

As in mammals, the pars distalis is irrigated by a hypophysial portal system which arises as a primary plexus on the median eminence. This plexus is drained by portal vessels that enter the pars distalis in the dorsocentral region and then break up into a secondary plexus. Blood from the pars distalis is drained directly into the cavernous sinus which communicates with the carotid and jugular veins. Blood is supplied to the primary portal plexus by the infundibular arteries. This portal system was described almost simultaneously by Green (1951) in the chicken, by Benoit and Assenmacher (1951) and Assenmacher (1952) in the duck, and by Wingstrand (1951) in several other species. The crucial demonstration of blood flow from the median eminence to the pars distalis was first accomplished by Assenmacher (1958) in the duck.

Since these studies, further interest has centered on the differential distribution of the portal vessels between the lobes of the pars distalis, and the most thorough study by Vitums et al. (1964) in *Zonotrichia leucophrys gambelii* has established several points: (1) There are distinct anterior and posterior primary capillary plexus corresponding with the anterior and posterior divisions of the median eminence. (2) These plexus are drained by separate anterior and posterior groups of portal vessels. (3) The anterior vessels are distributed mainly in the cephalic lobe of the pars distalis, while the posterior vessels supply the caudal lobe. A point-to-point distribution of the portal vessels has also been described in less detail for fifteen other species of birds by Dominic and Singh (1969). The degree of vascular separation is species dependent, and anterior and posterior capillary plexus are not distinct in the Cattle Egret (*Bubulcus ibis*), the Indian Roller (*Coracias benghalensis*), and the White-breasted Kingfisher (*Halcyon smyrnensis*). In the Japanese Quail, the portal vessels do not form separate groups beneath the median eminence before entering the pars distalis but pass directly as single vessels into the pituitary throughout its length. In most specimens, there is evidence for two portal plexus supplying the two lobes of the pars distalis. It must be emphasized that this distribution of portal vessels has not yet been shown to have any physiological significance.

Embryology

Morphogenesis

The adenohypophysis is formed exclusively by Rathke's pouch. At an early stage of development (6 days in the domestic fowl) there is subdivision of the tissues, the proliferation of which result in the different parts of the adult adenohypophysis: (1) The aboral and oral lobes give rise, respectively, to the caudal and cephalic lobes. (2) The lateral lobes form the pars tuberalis. (3) The contact zone with the brain remains as a single layer of cells and corresponds morphologically with the pars intermedia. Later these cells lose their contact with the pars nervosa and are included in the pars distalis.

The vasculogenesis of the pituitary has been studied in the Pekin duck by Assenmacher (1951, 1952). The hypophyseal portal system is established simultaneously with the proliferation of the oral and aboral lobes of the anlage. The primary plexus on the diencephalon appears first and is drained by venules that penetrate the proliferating cellular cords of the gland and are later transformed into the capillaries of the secondary plexus. A similar situation occurs in the chick embryo.

Cellular differentiation

The chronological appearance of the cell types has been studied in the chick embryo, but the results have varied with the methods and with the authors. Three kinds of staining methods have been used:

1. Tinctorial methods: Severinghaus, Altmann's fuchsin, Azan technique, Romeis' cresazan.
2. Silver staining according to Bodian.
3. Cytochemical techniques for glycoproteins [periodic acid-Schiff (PAS)], Gomori's paraldehyde fuchsin.

The first morphological signs of cellular differentiation appear in the cephalic lobe where the first acidophils are generally found on day 10 or 11. The first PAS-positive cells occur either on day 6 or on day 11. In the caudal lobe, differentiated cells appear much later, with acidophils being seen on either day 15 or days 18-19, and the first PAS-positive cells occurring only after hatching. The appearance of acidophils and of PAS-positive cells within the cephalic lobe is closely associated with the more precocious establishment of the vascular pattern in this zone. Electron microscopy should provide a better estimate of when secretory activity starts in the embryo, and recently Guedenet et al. (1970) have identified the first secretory granules in the cephalic lobe on day 8.

Tentative correlations have been made in the embryonic chick between morphological and functional differentiation of the adenohypophysis. Melanotropic hormone (MSH) activity has been demonstrated in the 5-day embryo and seems related to the appearance at the same time of the first strongly silver-positive cells. By using a sensitive *in vitro* method, thyrotropic activity was first found on day 7 of development. This precedes the appearance of PAS-positive cells in the cephalic lobe. Later, however, there are clear parallels between the evolution in both numbers and cytology of the PAS-positive cells and changes in thyroid activity. Although there is some evidence for the presence in the embryonic chick of gonadotropic and corticotropic activities, no cellular localization for these activities has been suggested.

Experimental embryology

Knowledge about the differentiation of the adenohypophysis has been greatly enhanced by the work: of Ferrand and Le Douarin with the embryonic chick. Their main findings may be summarized as follows. Before the 25 somite stage, the infundibular floor of the forebrain is necessary to induce the formation and maintenance of Rathke's pouch. These results confirm the work of Stein (1929, 1933) and Hillemann (1943). After this stage, however, the adenohypophyseal anlage can develop normally in the absence of the hypothalamus. Thus, the 3-day embryonic adenohypophysis is maintained as a coelomic graft for 14 days and during this period undergoes normal histological development with acidophilic and PAS-positive cells appearing at the same time as *in situ*. Two acidophilic cell types are distinguishable, and thyrotropic activity appears under these conditions. This latter finding corroborates previous results showing the independence of pituitary thyrotropic function from the hypothalamus in the chick embryo.

Ferrand's discoveries are of particular importance in view of the role previously assigned to pituitary vasculogenesis in the histological development of the gland. It now seems that during development the pituitary cells do not require specific substances carried from the infundibulum by the portal system. Some recent results of Ferrand (1971) indicate, moreover, that the vascularization exerts a favourable but only trophic effect; in chorioallantoic grafts on decapitated chick embryos, the new vascularization pattern stimulates the differentiation of PAS-positive cells in the caudal lobe.

MORPHOLOGICAL IDENTIFICATION OF CELL TYPES

Particular difficulties are encountered in identifying avian pituitary cell types because the cells are rather small and show only weak affinities for most dyes. As a first step, investigators used well-defined staining methods, such as those of Severinghaus, Bodian, Heidenhain, Azan, and Romeis, and were able to distinguish three cell classes: (1) large and deeply staining acidophilic cells (Al) in the caudal lobe, (2) small and slightly staining acidophils (A2) localized at the rostral end of the cephalic lobe, and (3) basophilic cells that were generally small and difficult to stain and that tended to lie in the cephalic lobe. Rather more refined analyses became possible when glycoprotein-staining methods were introduced into pituitary cytology. This second phase of research was started by Legait, who identified PAS-positive and AF-positive cells in the hen pituitary. Thereafter, the methods were developed further by Mikami (1958) for the cockerel and by Herlant et al. (1960) and Tixier-Vidal et al. (1962) for the Pekin duck. The third phase of cytological research has employed the electron microscope. Such studies were first performed in the hen, duck, and pigeon, and lately have been extended to include the chicken, White-crowned Sparrow, and Japanese Quail.

Technology

In spite of the constant technical progress, it often remains difficult to correlate the results obtained by different authors in the various species. This seems to result from the many variations that occur both with the techniques and with the stainability of pituitaries of different species. Before outlining the present stage of our knowledge, it seems useful, therefore, to recall some of the more suitable staining methods.

Fixatives

The identification of the cell types requires excellent preservation of the cytoplasmic structures and especially of the secretory granules, and the choice of an appropriate fixative is of paramount importance. The most commonly used are Bouin Hollande sublimate according to Herlant (1960), Zenker-formol, formol calcium, and formol sublimate. Acid fixatives such as Bouin or Susa cause modifications in the affinities of the cells for the stains, in spite of giving a good general appearance to the tissue. For ultrastructural studies, osmic acid and glutaraldehyde give excellent results.

Tinctorial methods

The classical trichrome methods have been used for a long time in avian pituitary cytology, but more recently some specialized polychromic methods have been introduced. One of the most commonly used is the alizarin blue tetrachrome of Herlant (1960), which has the advantage of differentiating five tinctorial types of cell—orange, rose, purple, light blue, and very dark blue. This method gives the best results if the tissues are fixed in Bouin-Hollande sublimate. Brookes' method (1967) gives excellent distinction between the various types of acidophils in birds. Matsuo has proposed a tetrachrome method involving orange G, acid fuchsin, methyl green, and acid violet. This technique stains the acidophils orange to red to purple, the basophils green, and the amphophils violet or purple; it invariably gives a deeply stained preparation. In the caudal lobe, the results correlate well with those of Herlant's tetrachrome, but in the cephalic lobe, the differentiation of the cell types is less clear.

Cytochemical methods

In contrast with the tinctorial methods, the cytochemical ones are easier to standardize, since the underlying chemical reactions are better understood. They offer more comparable results and have come into widespread use.

Glycoproteins

The PAS method plays a central role in pituitary cytology, its importance resulting from the glycoprotein nature of the gonadotropic and thyrotropic hormones. Often it is associated with acidic dyes such as orange G or naphthol yellow, which allow for the simultaneous detection of acidophils and glycoproteins. Matsuo frequently combines it with methyl blue.

In combination with a cationic dye such as alcian blue, it offers a useful means of distinguishing the acidic glycoprotein-containing cells, and several techniques have been proposed, all involving Alcian blue at low pH. By using the critical electrolyte concentration method of Scott and Dorling (1965), it has been possible to confirm in duck and Japanese Quail that the cells that retain Alcian blue are the richest in acidic groups. The PAS methods may also be combined with aldehyde thionin, but in our experience the results have been less satisfactory than with Alcian blue-PAS methods.

The technique of following PAS with lead hematoxylin is of particular value since it allows a deeply black-staining cell to be distinguished from the rose-coloured glycoprotein cells. Pretreatment of the sections with methasol blue or luxol fast blue before PAS can be useful in characterizing one glycoprotein cell type in the cephalic lobe. Two methods have been used to detect glycoproteins with the electron microscope; both appear successful.

Gomori's paraldehyde fuchsin

The cytochemical significance of this dye is not clear. Although somewhat less selective than PAS, it generally stains glycoprotein-containing cells and in its various forms has been applied to a number of species.

Amino acids

Disulfide groups have been detected in the duck pituitary with the method of Barnett and Seligman (1954), the strongest reaction being seen in a single cell type in the caudal lobe. Two cell types in the cephalic lobe of the Japanese Quail show a high tryptophan content when stained by Glenner's rosindol method.

Biogenic amines

In contrast with some mammals, no specific green or yellow fluorescence could be demonstrated by the Falck-Hillarp technique in the pituitary of the pigeon. This means that the cells contain neither catecholamines nor serotonin, although some cell types, notably those staining with PAS and AF, are capable of storing noradrenaline if it is administered exogenously. In the Japanese Quail, some cells in the cephalic lobe (PAS-positive, lead hematoxylin-positive) show an uptake of [^{3}H] serotonin *in vitro*.

Phospholipids

Phospholipids have been detected with Baker's acid hematin test in a caudal acidophil of the fowl pituitary. In the Japanese Quail, the same technique gives the strongest reaction in two cell types in the cephalic lobe and one in the caudal lobe.

Enzyme cytochemistry

The localization of acid phosphatase has been studied with the electron microscope in duck and quail pituitaries. It appears that even in normal birds, lysosomes are very abundant in the pituitary cells. A similar observation has been made with the embryonic chick pituitary. Acetylcholinesterase activity has been localized in PAS-positive cells of some bird pituitaries—White-crowned Sparrow, Brambling (*Fringilla montifringilla*), House Sparrow (*Passer domesticus*), European Starling (*Sturnus vulgaris*), and the canary, but not in the Japanese Quail, chicken, or pigeon.

Ultrastructural data—their importance as indicators of the functional state of pituitary cells

Two kinds of criteria are useful in the study of pituitary cells, those that allow for the differentiation and identification of the cell types, and those that are indicators of the functional state of the cells. Information from the electron microscope falls into the latter category and is of particular importance in birds because the pituitary cells are so small and often difficult to stain. Some cytoplasmic structures, of course, such as the Golgi apparatus and the rough endoplasmic reticulum (RER) can be examined only with the electron microscope. When dealing with cellular function we shall attach some importance to ultrastructural data. The identification of cell types with the electron microscope is not easy, since only black and white pictures are obtainable. Two criteria are useful, however; the mean size and electron density of the secretory granules and the features of the RER or ergastoplasm.

Tinctorial Affinities, Cytochemical Properties, and Ultrastructural Features

The tinctorial and cytochemical properties of the avian pituitary cells have been studied thoroughly in only a few species—the domestic fowl, both male and female, the pigeon, the Pekin duck, the Red Bishop, the quail, and the White-crowned Sparrow. In most, the affinities of the cells for the various stains are altered greatly by the physiological condition of the animal, and at first sight, therefore, the cell types seem to differ greatly among species. However, it is possible to find characteristic features for each cell type, provided they are studied at the time of maximum stainability. This invariably means comparing glands from birds in various physiological states. As in other vertebrates, some species are far better than others for morphological analyses, e.g., the Pekin duck.

A morphological analysis of the pituitary cell types necessitates some form of nomenclature. The International Committee recommended the use of a uniform and functional nomenclature, but this is still impossible in birds because of disagreement over the localization of several of the pituitary hormones. In this chapter, therefore, we shall continue to use Romeis' Greek-letter nomenclature as modified by Herlant. It is no better than any other system, but has been used widely for duck, pigeon, and quail pituitaries. Each cell type will be defined by its most typical tinctorial and cytochemical features. According to Herlant (1964), we shall distinguish two general classes of cells—those whose granules contain glycoproteins, and those whose granules contain simple proteins. Of the seven cell types recognized in the bird pituitary, only five can easily be assigned to these two classes. The remaining two cell types form a third class, characterized by having mixed affinities.

Proteinaceous cells (Acidophils, Serous cells)

α Cells (Caudal acidophils)

The α cells occur in the caudal lobe. They have the same properties as the α cells of other vertebrate groups and are selectively stained by all strongly acid dyes—orange G, erythrosine, naphthol yellow, and acid fuchsin. They contain disulfide groups due to a high concentration of cysteine and also contain phospholipids and lipoproteins. They are PAS-negative except after permanganate oxidation, when they appear pale rose due to the appearance of some hydroxyl groups. They are generally larger than any other avian cell type and show a strong polarity.

The α cells are among the types included in the classic Al cells of Rahn and Painter. In fact, after staining with Heidenhain's azan, or with Matsuo's tetrachrome, it is possible to distinguish two types of Al cells. With Herlant's tetrachrome and with PAS methods these cells can be distinguished as the acidophilic α cells and the γ glycoprotein cells. α Cells are far less numerous in birds than in mammals.

Under the electron microscope, a cells have been recognized in the duck and in the White-crowned Sparrow, in which they seem to have the same features. The granules are uniform in shape and size (250-300 nm), but not in density. The endoplasmic reticulum appears as a series of flattened sacs often dilated by large and irregular vacuoles.

η Cells (Cephalic acidophils, Erythrosinophilic cells)

These cells are characterized by their selective affinity for erythrosin in the tetrachrome stains of Herlant and of Brooke. They can also be distinguished from the a cells because they lie in the rostroventral region of the cephalic lobe, are irregular or polyhedric in shape, contain much less phospholipid, and are generally less easily stained. They seem to correspond with the A2 cells of Rahn and Painter.

Electron microscope observations of these cells have been made for the duck, the pigeon, and the White-crowned Sparrow. They are characterized by their polymorphic secretory granules, the size and density of which vary with the species and with the functional state of the gland. As with the α cells, the rough endoplasmic reticulum appears as flattened or dilated sacs.

Glycoprotein-containing cells

β Cells (Cephalic PAS-positive cells)

These cells of the cephalic lobe are the most strongly PAS-positive cells in the gland. This is especially so in seasonally breeding species in which they are more easily distinguished than in domesticated birds such as the chicken and feral pigeon. They contain some phospholipid, and in the Japanese Quail, a high level of tryptophan. Typical acid dyes scarcely stain the β cells. With Herlant's tetrachrome, identification is sometimes difficult, since their affinities for aniline blue and alizarin blue depend upon the glycoprotein content, and they may be stained from purple to blue. The same difficulty occurs with the classic trichrome stains. Characteristically these cells are small and narrow with a strong polarity and could correspond well with the

V cells described in the chicken by Mikami (1958). They belong to the basophil or amphophil class of pituitary cells.

At the ultrastructural level they have been descried in the duck and the Japanese Quail. Typically, they contain numerous round, uniformly dense granules of diameter 200-250 nm. The rough endoplasmic reticulum generally consists of scattered, flattened cisternae with no dilations. These features are similar to the type A gonadotropic cell found in both pituitary lobes of the White-crowned Sparrow by Mikami et al. (1969).

δ Cells (Alcian blue-positive cells)

This cell type occurs in both lobes (duck, Japanese Quail, guinea fowl, pigeon, and Red Bishop), although it is often smaller in the caudal lobe. They are PAS-positive, though less so than the β cells, and after Alcian blue-PAS are selectively stained a blue colour. In the duck, they stain selectively with AF, but in several other species the aldehyde-fuchsin is also taken up by β cells. Up to now, therefore, the best criterion for their identification seems their affinity for Alcian Blue. Unfortunately, these cells are often chromophobic and in some species only acquire their characteristic affinities under limited conditions. Personal experience suggests that they appear best in the duck pituitary. With tinctorial methods, they stain with aniline blue and light green and belong, therefore, to the class of basophils. Their rounded shape with a central nucleus and a general paucity of granulations also helps to distinguish them from other cell types. In some species (Japanese Quail, ovariectomized hen and cockerel) they contain large PAS-positive droplets.

Under the electron microscope these cells are characterized by having a few small granules (50, 150, 200 nm) and by the great development of dilated and rounded vacuoles within the cisternae of the rough endoplasmic reticulum. Glycoproteins are localized exclusively in the secretory granules, and this probably explains their low affinity for stains in the light microscope. In the Japanese Quail the rounded cisternae contain large PAS-positive inclusions.

γ Cells (Glycoprotein-containing acidophils)

The γ cells, which occur in the caudal lobe, are a mixed cell type combining an affinity for PAS with some characteristics of acidophilic cells. The glycoprotein content changes considerably with the functional state of the gland, and such variation probably explains the variability in their reaction to the tetrachrome stains. Sometimes they appear as acidophils, sometimes as cyanophils or amphophils.

With the classic trichrome techniques, such as Heidenhain's azan, they can easily be confused with α cells. Also in common with the α cells, they contain S—S groups and phospholipids as well as having the same general shape. For these reasons they are considered together with the α cells to represent the Al cell type of Rahn and Painter. In the majority of species they are somewhat more numerous than the α cells, although this does not apply to the Japanese Quail. In some species, particularly in the Red Bishop, they sometimes show a high affinity for lead hematoxylin.

γ Cells have been distinguished with the electron microscope in the duck and Japanese Quail. They have dense, round secretory granules of diameter 300-400 nm. The rough endoplasmic reticulum is well developed and consists of flat cisternae lined with numerous ribosomes. In the pigeon, the granules are polymorphic and the reticulum often dilated.

Mixed cell types

ε Cells

These cells are evident only under conditions associated with a decrease in adrenocortical activity and are difficult to distinguish in normal birds. They are restricted to the center of the cephalic lobe. The cytoplasm usually has a flocculent appearance due to the small size of the secretory granules. They are slightly acidophilic and PAS-positive and have a distinct, although not strong, affinity for lead hematoxylin. With tetrachrome staining methods they react best to Brooke's technique, which stains them a gray-rose colour set against a general green background. These cells could well correspond with some of the "V cells" in the chicken.

Ultrastructurally, they can be distinguished in the duck after metipirone treatment and in the chicken and White-crowned Sparrow after adrenalectomy, when the rough endoplasmic reticulum swells significantly. The secretory granules are dense and small (mean diameter 150 nm) in the duck and chicken. They are a little larger (220 nm) in the White-crowned Sparrow and apparently show two distinct phases of low and high electron density.

κ Cells (Lead-hematoxylin cells)

In contrast with the ε cell, the κ cell is highly stainable, but it remains difficult to classify. With Herlant's tetrachrome, the cells are dark blue and could thus be considered as basophils. Indeed, they are often well stained with toluidine blue at pH 4.5, sometimes with a slight metachromasia. Kappa cells are also strongly stained by iron hematoxylin and lead hematoxylin. In the Japanese Quail, the cells contain high levels of both phospholipid and tryptophan. Their low glycoprotein content distinguishes them from β cells. There is some species variation in their localization within the gland. For example, in the Pekin duck, they are scarce and restricted to the cephalic lobe, while in the House Sparrow and Red Bishop they are more abundant but still localized within the cephalic lobe. In the pigeon and Japanese Quail, they occur throughout the gland.

Electron microscope observations of κ cells have been described in the duck, pigeon, and Japanese Quail. In all, they contain round dense granules, which are the largest in the gland (400-500 nm). The endoplasmic reticulum is often well developed and appears as flattened sacs lined with many ribosomes.

In summary, this review confirms the pituitary cell pattern originally proposed by Rahn, Painter, Payne, and Wingstrand, but extends it from 3 to 7 cell types. These cell types are not distributed evenly through the gland, although the degree of segregation is not as pronounced as in some other vertebrates, notably the fishes. This becomes clear if one compares serial sections cut either in the transverse or sagittal plane. The mixed zone between the caudal and cephalic lobes occupies a full third of the gland and, moreover, does not lie in a fixed position but can vary depending on the physiological state of the bird.

Cytological Localization of the Pituitary Hormones

Two facts appear reasonably well established. First, histological methods demonstrate the presence of seven cell types in the avian pituitary. Second, the gland seems to secrete seven hormones—two protein hormones [growth hormone (somatotropin, STH) and prolactin]; three glycoprotein hormones [thyrotropin (TSH)] and two gonadotropins (GTH) [luteinizing hormone (LH) and follicle-stimulating hormone (FSH)]; and two polypeptide hormones [adrenocorticotropin (ACTH) and melanotropic hormone (MSH)]. The aim of the histophysiological study of pituitary cells is to correlate each morphological cell type with one hormone. This presumes, of course, that each cell type secretes one hormone, a hypothesis that is not altogether fully sustained.

Currently, two methods are available to establish the nature of the hormone secreted by a particular cell type. The best is the immunocytochemical technique, which directly localizes the hormone within its cell. As yet this method has not been applied to birds, but the recent purification of some of the avian hormones suggests it should be introduced fairly soon. The alternative approach is experimental and offers indirect evidence for the localization of a particular hormone. Invariably this technique involves altering the activity of the different pituitary target organs, thereby changing the rate of secretion of a particular tropic hormone. Birds from a wide number of physiological conditions must be studied, and the method requires a preliminary knowledge of the pituitary cytology. The method leads finally to tentative conclusions regarding the localization of the various tropic hormones.

Thyrotropic Activity

Thyroidectomy

The thyrotropic cells have been identified following thyroidectomy (chicken, Pekin duck, Japanese Quail, White-crowned Sparrow) and treatment with antithyroid drugs (chicken, duck, Red Bishop, Japanese Quail). All authors agree that thyroidectomy strongly stimulates one PAS-positive cell type which corresponds with the δ, or Alcian blue-positive cell described earlier and belongs to the basophilic class of the avian pituitary. After thyroidectomy, the cells are greatly hypertrophied and become hyalinized, less easily stained, and often vacuolized. Electron microscope pictures of these thyroidectomy cells are very similar in the chicken, White-crowned Sparrow, and Japanese Quail, with the endoplasmic reticulum being greatly developed so as to occupy the greater part of the cytoplasm. There is also a loss of secretory granules, the few remaining being very small (100 nm) and scattered along the cell membrane. The Golgi zone is as large as the nucleus and contains numerous small granules. In both the light and electron microscope it is possible to follow the progressive changes of the normal δ cell into a thyroidectomy cell.

Opinions differ over the localization of this thyrotropic cell type. In the chicken and White-crowned Sparrow, Mikami observed thyroidectomy cells only in the cephalic lobe and thus restricts TSH-producing cells to this zone of the pituitary. In the Japanese Quail, we also saw thyroidectomy cells in the cephalic lobe 13 and 40 days after the operation. Similarly, in the duck and the Red Bishop, after 1 month of treatment with thiourea, the δ cells were stimulated only in the cephalic lobe. But, in Japanese Quail, the same treatment led to an increase in δ cells in both lobes, and in ducks killed 14 months after thyroidectomy, there were hypertrophied δ cells in both lobes of the gland also. These conflicting results may be resolved if it is assumed that thyrotropic cells exist in both lobes, but that in the cephalic region they are more sensitive to modifications in the level of thyroidal hormones or more prone to become "thyroidectomy cells." Indeed, one must notice that small thyrotropic cells always persist in the cephalic lobe and that the thyroidectomy cell represents a maximum stage of stimulation. This hypothesis is not in conflict with Mikami's findings that TSH concentration in the normal chicken pituitary is higher in the cephalic lobe and that only in the cephalic lobe is there an increase in TSH content following thyroidectomy. Similarly, in the chick embryo, thyrotropic cells appear first at the rostral end of the cephalic lobe and only after hatching do they appear in the caudal lobe. TSH activity is exclusively found in the cephalic lobe of 1-day-old chick pituitaries. The only other cells modified by thyroidectomy are the caudal acidophils, which undergo a marked degranulation and are reduced in volume and in number. This is considered a secondary phenomenon, the endocrine significance of which is not understood.

Effect of treatment with thyroid hormones

After injection of thyroxine, the δ cells became highly chromophilic and stain deeply with Alcian blue. This effect occurs in both lobes of the gland. In thyroidectomized quail, injections of thyroxine induce repression of the thyrotropic cells and the appearance in the cytoplasm of numerous dense bodies with a high level of acid phosphatase.

Several arguments, therefore, support the view that TSH is produced by the Alcian blue-positive, or δ, cells. These may be localized either in the cephalic or in both lobes, but the activity is greatest in the cephalic lobe. The properties of this cell are similar to the thyrotropic cell in other vertebrates.

Gonadotropic Activity

Information from three sources is generally used to identify the gonadotropin-producing cells: (1) the changes occurring during a natural or artificially induced breeding cycle, (2) the effects of castration, and (3) the effects of treatment with sex steroid hormones.

Reproductive cycle

The pronounced reproductive cycle of most wild birds, together with the ease of modifying this cycle experimentally, makes such species more attractive for cytological analysis than domesticated forms that are continually in full sexual activity. Nevertheless, in both the domestic chicken and the pigeon the pituitary undergoes cytological changes that have been studied relative to the ovarian cycle. The hormonal control of the ovary is complex, contributing to difficulties in interpretation. The clearest evidence concerns PAS-positive cells that display their greatest activity in laying hens and regress during broodiness. They seem to be involved in a gonadotropic function but may also have a role in TSH production. In addition, the caudal acidophils undergo modifications that could also be correlated with the ovarian cycle.

Changes in pituitary cytology through the reproductive cycle have been analyzed in the male of several species—the Pekin duck, the Red Bishop, the Japanese Quail, and the White-crowned Sparrow. In the duck, Red Bishop, and Japanese Quail, the most striking changes concern the β or PAS-positive cells of the cephalic lobe. This cell type is almost absent during sexual quiescence. Its appearance and numerical development is, however, closely tied both to the testicular cycle and to the gonadotropic content of the pituitary. A second glycoprotein-containing cell, the γ cell, is also implicated in the sexual cycle. In the duck, the γ cells are highly condensed out of the breeding season but undergo degranulation and activation of the Golgi zone when testicular growth commences. Pronounced changes also occur in the γ cells of the Red Bishop and Japanese Quail during testicular development, and the cells become so degranulated as to be difficult to recognize with the light microscope. They reappear during testicular regression.

In the White-crowned Sparrow, Matsuo et al. (1960) have observed a good correlation between the reproductive cycle of both males and females and changes in the "light and deep basophils," which occur in both lobes. These PAS-positive cells contain acetylcholinesterase, and there is a parallel between the time course of AChE activity in the pars distalis and the growth of the testis.

The δ cells of the Pekin duck also show an annual cycle, but it differs in phase from that of the testis; they are in a resting state in March-April when the testis is undergoing maximum growth and when the thyroid is only weakly active. In this species, therefore, we do not consider the δ cells as gonadotropic. In artificially photostimulated Japanese Quail, Red Bishop, and ducks, this dissociation between the β and δ cells is much less evident. There is, of course, positive evidence that the δ cells in the Red Bishop and Japanese Quail have, as in the duck, a thyrotropic function. In the White-crowned Sparrow, the light basophils, which may be homologous to our δ cells, are also stimulated in phase with the testis, but it is not clear that these cells have a thyrotropic function.

Castration

Castration in photosensitive birds

The effects of castration are closely related to the photoperiod. In Japanese Quail, the pituitary cytology is not modified if the castrated birds are held on short daily photoperiods, a fact corroborated by the small changes observed in both pituitary gonadotropin and plasma LH content. The same appears true for the White-crowned Sparrow although in the electron microscope one cell type (type B gonadotropic) was found to be stimulated; but the AChE cells are not modified. In the castrated duck held on short days, the cephalic β cells are not modified.

In contrast, there are marked cytological changes if the birds are killed after a period of photostimulation. Such a treatment also invariably leads to pituitary hypertrophy and to an increase in pituitary and plasma gonadotropins. At the cytological level, two types of result have been reported—those of our own on the duck and Japanese Quail, and those of Matsuo and Mikami with the White-crowned Sparrow.

Three cell types are stimulated in the duck pituitary. Two of these—the β cell of the cephalic lobe and the γ cell of the caudal lobe—show an increased abundance and become hypertrophied and degranulated. The endoplasmic reticulum remains as flattened cisternae and there is no vacuolization of the cytoplasm. These effects agree well with the changes found during the annual reproductive cycle and confirm the view that the β and γ cells produce gonadotropic hormones. The third cell to be stimulated is the δ cell, previously identified as the TSH-producing cell. This cell also hypertrophies but in contrast with the other two types, also shows a high degree of vacuolization. Under the electron microscope, it appears rather similar to the thyroidectomy cell. One difference, however, is that the stimulated δ cells occur in both pituitary lobes. Castration seems, then, to stimulate thyrotropic activity, a finding supported by physiological experiments.

Much the same pattern of changes is observed in photostimulated castrated Japanese Quail, although the γ cells become difficult to interpret because of their low stainability. The δ cell of the castrated Japanese Quail differs from the thyroidectomy cell in two respects: (1) there is equal stimulation of the cell in both lobes, and (2) the cells contain numerous lysosomes. This suggests an intracellular degradation of the hormone and implies a lower rate of TSH release in the castrate than in the thyroidectomized bird. In fact, the stimulation of thyroid activity in the Japanese Quail following castration is much less pronounced than in the duck.

In the White-crowned Sparrow after castration and photostimulation, Mikami et al. (1969) described the stimulation of two cell types occurring in both lobes of the pituitary. One of them, the type A gonadotropic cell, hypertrophies and shows a well developed endoplasmic reticulum with dilated cavities and a large Golgi zone. The other cell type, type B gonadotropic cell, is characterized by a vesiculated cytoplasm containing vacuoles without conspicuous limiting membranes and a few dense secretory granules (180-350 nm). Curiously, this type B gonadotropic cell is stimulated in unilaterally gonadectomized and photostimulated birds and in 10-day castrates that have not been photostimulated. In photostimulated castrates of 14 and 65 days' treatment, the cell type is "rarely observed." The AChE cells are hypertrophied when castration is followed by photostimulation. They seem to represent one cell type only, occurring in both lobes.

If one tries to compare electron microscope pictures of Mikami et al. (1969) from the White-crowned Sparrow with ours from duck and Japanese Quail, some analogies do appear:

1. In normal sexually active males, type A gonadotropic cell is quite similar to our β cell.
2. In castrated photostimulated males, type A gonadotropic cell appears similar to our δ cell.
3. In the three species, thyroidectomy cells show the same ultrastructural features.
4. We can find no analogy between Mikami's type B gonadotropic cell and our γ cells.

Castration in chicken and pigeon

Castration is the only method in these species of identifying the gonadotropic cells. In both it causes hypertrophy of the PAS- and AF-positive cells (pigeon; hen; cockerel). The castration cells are located in both lobes and Mikami (1969) considers that in each of them two cell types are involved as in the White-crowned Sparrow. In the pigeon, on the contrary, two gonadotropic cells are stimulated, the γ cells in the caudal lobe, and the β cells in the cephalic lobe; the δ cells, considered thyrotropic, are stimulated in both lobes. An increase of the thyrotropic potency of the pituitary has been previously observed in the castrated pigeon.

There is no simple explanation for the differences between the two schools on this problem, although we feel that the explanation probably lies less in fact and more in interpretation. At the present moment, the two views can be summarized as follows:

1. The studies in the duck and Japanese Quail suggest that the gonadotropins are produced by the γ cells localized in the caudal lobe and the β cells in the cephalic lobe. Castration also stimulates the δ cells, suggesting a relationship between thyrotropic and gonadotropic functions.
2. In the White-crowned Sparrow and chicken, the gonadotropic activity is localized in two cell types occurring in both the cephalic and caudal lobes. The thyrotropic cells are small in size and number, suggesting that gonadotropic and thyrotropic functions are not interrelated.

Effects of sex steroid and of thyroxine injections

Testosterone

The effects of testosterone propionate have been studied in the Pekin duck with both the light and the electron microscope. Both intact and castrated ducks have been used in combination with short and long days. Testosterone induced a storage of secretory granules in several types of pituitary cells, including the two gonadotropic cells, β and γ, and the thyrotropic δ cells. It partially inhibited the effects of castration plus photostimulation. These findings imply that testosterone has restrained the secretion of thyrotropin as well as the gonadotropins. There is good evidence from bio- and immunoassays that gonadotropin secretion is blocked by testosterone, but the effects on TSH secretion are still unclear. Prolactin secretion is also inhibited by testosterone, and a storage of the secretory granules has been observed in the prolactin cells.

Thyroxine

The effects of thyroxine injections have been studied in intact and castrated male Japanese Quail maintained on long days. In both cases, thyroxine induced a storage of granules and a regression of both the β and δ cells, but the active dose is higher in the castrated than in the intact male. Numerous lysosomes appear in the two cell types.

In conclusion, the effects of testosterone and thyroxine do not aid the interpretation further. If one assumes that the three cell types—γ, β, and δ—each produce a different hormone, then one must conclude that they possess common receptors for testosterone and for thyroxine, and that the regulation of the pituitary-thyroid and pituitary-gonad axes are closely related. Another possibility may be that a common subunit exists in the avian thyrotropic and gonadotropic hormones, the synthesis of which is regulated by a common mechanism.

LH and FSH localization

In the Pekin duck it is thought that the γ cells produce the LH (or ICSH) activity, while FSH arises from the β cells. Direct proof of this is unavailable, but in the duck, the two gonadotropic cells develop asynchronously during the annual cycle. The γ cells show a maximum in March but are regressed in April-May when the testes reach their greatest weight. Gamier has shown that plasma testosterone reaches its peak in March. In October, there is another peak of plasma testosterone, although the testicular weight is minimal. At that time, the caudal lobe γ cells show some activity, while the β cells have almost disappeared. Castration at this time activates only the γ cells. The conclusion is that the γ cells are more closely related to testosterone production, and the β cells to testicular weight and development of the seminiferous tubules. The distribution of an LH activity in the caudal lobe and an FSH activity in the cephalic lobe has also been proposed for the day-old chick by Brasch and Betz (1971), who transplanted various regions of the pituitary onto the chorioallantoic membrane of partially decapitated chick embryos. Subsequently, the histology of the embryonic gonad was analyzed.

Prolactin

Prolactin does not have a target organ that can be ablated to stimulate hormone secretion; thus, the identification of the prolactin-producing cells largely rests on correlating cytological changes with the prolactin content of the pituitary. This content is raised during broodiness and care of the young in the hen, the pigeon, the California Gull (*Larus californicus*), the Ring-necked Pheasant and the turkey. Nakajo and Tanaka (1956) have also shown that in nonbroody hens prolactin potency is greatest in the cephalic lobe, whereas the converse is true of broody birds. Prolactin content appears to follow an annual cycle concomitant with testicular development and is increased by experimental photostimulation.

Using the light microscope, the cytological changes occurring at broodiness have been studied in the pigeon and the hen. In the latter, broodiness is accompanied by regression of the basophils and development of the cephalic (α2) and caudal (α1) acidophils. The former seem most closely related to broodiness. Broody or lactating pigeons show development of the cephalic acidophils that are typical erythrosinophils, as are the prolactin cells of several vertebrate classes. The caudal acidophils showed no changes in the pigeon. Frequently, however, clumps of η cells are seen in the caudal lobe of broody or lactating pigeons, showing again how the two structures of the avian pituitary may become mixed when a strong stimulation is applied. The erythrosinophilic or η cells have been found in the cephalic lobe of several other species, including the duck, the Red Bishop, the guinea fowl and *Streptopelia* sp., the Japanese Quail, and the White- crowned Sparrow. In the duck and quail, their changes have been related to pituitary prolactin content, but this does not apply to the other species.

With the electron microscope, the prolactin cells have been identified and related to broodiness in the turkey, hen, and pigeon. They have also been identified in cultured duck pituitaries that synthesize and release prolactin at a constant rate. The prolactin cells of the broody turkey look like rat prolactin cells with large, round, secretory granules and numerous flat ergastoplasmic cisternae. In broody hens and pigeons, they contain small polymorphic granules and have a much enlarged endoplasmic reticulum. Similar results were found with the duck pituitary, the granule size varying with the phase of cellular secretion. This cell type was stimulated by permanent light and degranulated after reserpine injections, which was consistent with the changes in pituitary prolactin content.

Mikami et al. (1969) tentatively identified a prolactin cell in *Zonotrichia leucophrys gambelii*. It was mainly localized in the cephalic lobe and had characteristically larve oval or irregularly shaped granules. The cells were more numerous and active in the photo-stimulated bird.

Corticotropin

Corticotropic cells are most easily identified following adrenalectomy, but in birds this operation is difficult and often lethal. It was achieved by Mikami (1958), who was the first to assign ACTH secretion in the chicken to a cell type located in the cephalic lobe—the so-called V cells. These increased in both number and size and were arranged in palisade layers throughout the cephalic lobe. They contained faint granules that lost their PAS- and AF-positive character. The changes were maximal after 48 hours and were prevented by injections of DOCA. The same observations have been made in the adrenalectomized White-crowned Sparrow. The ultrastructure of the pituitary adrenalectomy cells has been described for both species by Mikami (1969). Initially, the cells contained large secretory granules (220 nm) of both a low and a high electron density. Within 24-48 hours after adrenalectomy, the dense granules tended to disappear while the less dense granules became more numerous and lost their density.

Finally, they appeared as vacuoles surrounded by the remnants of the granular membrane. In the well developed adrenalectomy cell, these vacuoles occupy the entire cytoplasm and the endoplasmic reticulum, and Golgi apparatus degenerates—an odd response in a highly stimulated cell.

Metapirone seems to act in birds as in other vertebrates by inhibiting corticosteroid synthesis, thus increasing ACTH secretion. Its effects on pituitary cytology are very clear in the duck and Japanese Quail, with one cell type lying in the center of the cephalic lobe being stimulated and showing the characteristics of an adrenalectomy cell. In our classification this cell corresponds with the ε cell, being defined by its slight affinity for PAS and lead hematoxylin. It appears to be homologous to Mikami's V cell. Under the electron microscope, differences are apparent between ACTH cells in the duck treated with metapirone and in the adrenalectomized chicken and sparrow. The secretory granules in the duck are always dense, small (100-200 nm) and scarce, and lie along the cell membrane. The endoplasmic reticulum appears as parallel flat cisternae with small irregular vesicles or polyribosomes, and the Golgi zone is well developed.

Melanotropin (MSH)

The bird pituitary lacks a pars intermedia, but MSH activity occurs in the pars distalis, apparently being somewhat greater in the cephalic lobe. The first evidence for the cell type responsible for this activity came from the duck, where a cell was found that stained strongly with Herlant's tetrachrome and with lead hematoxylin. This cell type, the so-called κ cell, is restricted to the cephalic lobe. It bore affinities with a cell present in the fish pars intermedia and in the pars distalis of mammals (man and the pangolin), which also lack an intermediate lobe. In all avian pituitaries so far examined with these two staining methods a κ cell has been found.

The hypophysis that MSH derives from the κ cells has been recently examined further by Tougard (1972). There is a clear relationship between the number of κ cells and pituitary MSH activity in two breeds of duck, two pigeons, and two strains of Japanese Quail. In the ducks and pigeons, the number of κ cells was also correlated with feather pigmentation. While this evidence suggests MSH is produced by the κ cell, direct evidence is lacking. It is not helped by our ignorance of the role of MSH in birds. On several occasions we have reported strong changes in κ cell morphology, but these could not be correlated with a specific physiological event.

Growth Hormone or Somatotropic Cells

The caudal acidophils, or α cells, of the avian pituitary have the same properties as the α cell series in other vertebrates. In mammals, there is good evidence that the α cell secretes growth hormone, but in birds little is known of the existence or physiology of this hormone. Some analogies, however, imply that the avian α cell might also be the source of growth hormone. They are numerous and active in young ducks, Japanese Quail, and pigeons and very much less obvious in older animals. Furthermore, they are degranulated and appear regressed in thyroidectomized birds (chicken, duck), but reappear following treatment with thyroxine. Similar changes occur in the rat, where the relation between thyroid activity and somatotropin secretion has been carefully investigated. The α cells of the duck are also strongly affected by other treatments such as castration, long days, and reserpine. Such variations are consistent with the intervention of growth hormone in a wide range of metabolic processes.

Chemistry and Physiology of the Hormones

Gonadotropins (FSH and LH)

During the 1930s and early 1940s, many investigators showed gonadotropic activity in avian pituitary extracts, but whether this activity was associated with one or two hormones remained unclear. In fact, some results of Fraps et al. (1947) had suggested the existence of an LH and an FSH, but this was proven only when Stockell Hartree and Cunningham (1969) isolated two hormone fractions from chicken pituitaries, using as bioassays the standard mammalian techniques. These findings, which have greatly influenced this section, must lead to a reappraisal of much of the earlier work concerning the avian gonadotropins.

The problem still remains, of course, as to whether physiologically the two hormones may act together in some form of gonadotropic complex. This remains unresolved, although it seems unlikely, especially as Imai and Nalbandov (1971) report differential ratios of FSH:LH activity in the plasma of domestic hens at different periods of the laying cycle.

Assay

Without specific assays, progress into understanding the nature, number, and physiology of the gonadotropins is made immeasurably more difficult, and it is therefore the more unfortunate that most of the assays used hitherto show one or more grave defects.

Bioassays in mammals

The acceptable specific assays for mammalian LH and FSH are, respectively, the ovarian ascorbic acid depletion method (OAAD) in rats and the ovarian augmentation assay in rats or mice. This specificity seems also to apply to the two chicken gonadotropin fractions. Indeed, much of the basis for claiming the existence of chicken LH and FSH rests on the fact that of the two gonadotropin fractions isolated, one was active only in the Parlow assay, the other in the Steelman-Pohley assay. Both assay methods, however, are extremely insensitive to the avian hormones. Even with the OAAD assay it is usually impossible to assay the hormone present in a single pituitary gland of a hen if a factorial design (2+2) of standards and unknowns is used. For estimates of plasma LH, a dose of 3 ml/rat had to be used by Nelson et al. (1965). The FSH assays are more insensitive, Imai' and Nalbandov (1971) using doses equivalent to two and four hen pituitaries in a (2+2) design.

While the FSH assay seems specific, problems have appeared recently with the OAAD assay because of its responsiveness to arginine vasotocin, one of the natural neurohypophyseal octapeptides in birds. Jackson and Nalbandov (1969) report that vasotocin also occurs in high concentration within the chicken adenohypophysis and that estimates of OAAD activity in simple pituitary extracts measure both vasotocin and LH. This appears to be a serious, although surmountable, problem since vasotocin can be destroyed by sodium thioglycolate treatment, or separated from LH by dialysis or chromatography on Sephadex. Nevertheless, the finding raises problems over the validity of published data on LH levels in the chicken pituitary. Plasma estimates may be more accurate, for although vasotocin levels increase at oviposition, the amounts present should not affect the assay. In summary, the OAAD method seems acceptable for avian LH only if vasotocin is removed from any extract.

Bioassays in birds

The first valid assay depended upon the ability of gonadotropins to stimulate testicular growth in young domestic fowl chicks. A similar endpoint can be used in young turkey poults or in 14-day-old Japanese Quail reared under a nonstimulatory photoperiod. With these methods, however, the gain in sensitivity over the mammalian assays is marginal, and the methods have been largely superseded by an assay that measures the testicular growth response in terms of uptake of radioactive phosphorus.

As a bioassay the ^{32}P-uptake method is quite acceptable. The responses to mammalian gonadotropins (NIH-LH and FSH) remain stable over long periods, and adequate assay precision is obtainable with 6-8 chicks in each group and a typical factorial design. Again, the index of precision remains stable, as may be judged from three separate assay periods—in Pullman, where $\lambda = 0.154 \pm 0.006$ (SEM) (57); in Leeds, $\lambda = 0.219 \pm 0.010$ (43); and in Bangor, $\lambda = 0.228 \pm 0.014$ (42). The assay precision seems dependent on the strain of chick used. The method is sensitive to about 0.02 of a chicken pituitary (0.2 mg fresh weight) or to 0.2-0.4 of a pituitary from a Japanese Quail or White-crowned Sparrow. In practice, this means that single quail or sparrow pituitaries are assayable in a (2 + 1) design and that more complete potency estimates are possible with four or five pituitaries. Other advantages of the method include the ready availability, cheapness, and uniformity of 1-day-old chicks and the ability to process up to two hundred in a single assay.

However, the ^{32}P-uptake technique has one major drawback that greatly limits its usefulness; it responds equally well to both chicken LH and FSH and must therefore be considered as a total gonadotropin bioassay. According to Burns (1969), the method can be made specific for avian LH by treating pituitary extracts with neuraminidase, which selectively destroys FSH. This conclusion was based on the effects of

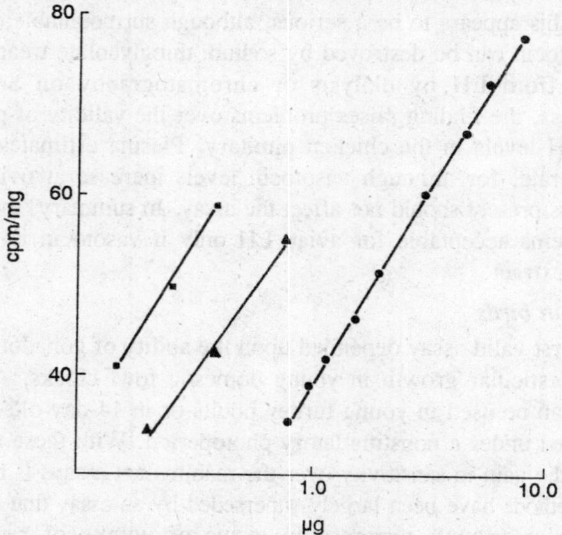

Fig. 1.1. Dose-response curves for gonadotrophins in the ^{32}P-uptake assay.

the enzyme on ovine FSH and on bovine LH. Unfortunately, our own results suggest that both chicken gonadotropins are equally susceptible to neuroaminidase action. The assay is unaffected by arginine vasotocin. Its value has lain primarily in its sensitivity and thus its use in measuring the hormone content in small birds. It has also enabled the hypothalamic gonadotropin-releasing factors to be characterized and assayed, the OAAD method being unusable for this because of vasotocin contamination of the hypothalamic extracts. The weaver-finch assay for LH may be specific, but as yet it has not been tested with purified avian hormones.

Immunoassays

The last seven years have seen the introduction of radioimmunoassay methods for the measurement of many hormones, and assays are now available for human, ovine, bovine, simian, and rodent LH and FSH, as well as for human chorionic gonadotropin. The advantages of immunoassay are a vast increase in sensitivity, allowing for the measurement of hormones in plasma, a more accurate estimate of potency, a stable and reproducible system, and the ability to measure many samples at once. Against this must be considered the problem of specificity. This includes both the specificity of an assay toward any single pituitary hormone, and also whether the system is measuring the same substance as the bioassay. Disparities, between bio- and immunoestimates of potency are quite common, and the problem remains central to all immunoassays.

A major technical problem is that most gonadotropin immunoassays have proved species specific; for example, antiovine LH does not cross react with chicken pituitary extracts when tested in an Ouchterlony plate, nor is it capable of blocking their biological activity. A weak cross reaction was observed, however, between chicken LH and antihuman LH in a hemagglutination inhibition system. Bagshawe et al. (1968) have reported a stronger cross reaction in an HCG radioimmunoassay with plasma from laying hens, and this reaction was used to estimate the circulating "LH" levels in the Greenfinch (*Carduelis chloris*) and the House Sparrow. No real evidence exists as to whether the assay is measuring avian LH, but the cross reactions are remarkable. Follett et al. (1972a) were unable to find an anti-HCG or antihuman LH serum that showed more than a trace ability to bind labeled chicken LH. The fact that different antisera to the same hormone show varied specificity is well established, and this is most probably the explanation for Bagshawe's results. Such cross reactions can be of great value, as exemplified by one antiovine LH serum produced by Niswender and Midgley that can be used to measure LH in sheep, rat, hamster, vole, and monkeys. In all cases, however, the assay has been validated with purified or semipurified hormones. The conclusions seem clear for avian physiologists; homologous assays using antisera and purified hormones from the species under study are the ideal; failing that an assay based on antibodies against the hormone from another species (heterologous assay) is acceptable if it can be validated. Both approaches became feasible when Stockell Hartree and Cunningham (1969) purified chicken LH and FSH, and a homologous assay for avian LH has now been developed.

The assay was developed with LH extracted from broiler chicken pituitaries. The antisera produced in rabbits had the ability to inhibit the biological activity of avian but not ovine LH *in vivo*. More purified LH fractions were labeled with radioiodine (^{125}I). The label bound to the antibody (antiavian LH) was separated from the unbound or free label by precipitation with an antirabbit γ-globulin serum. When increasing amounts of unlabeled chicken LH were added to this system, the binding of the LH-^{125}I was progressively inhibited, thus producing a standard curve. Chicken FSH appears to cross react only slightly in the assay. It was also concluded after a series of physiological

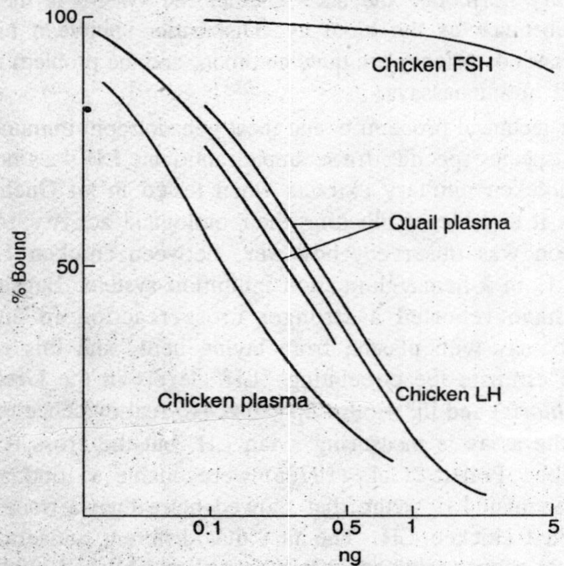

Fig. 1.2. Typical dose-response curves obtained in the avian LH radioimmunoassay.

experiments that the assay was not measuring significantly the other glycoprotein hormone TSH. Pituitary extracts and plasma from both the chicken and the quail cross react fully in the assay and the immunoreactive LH (IR-LH) level is measurable in the 2.5-200 μl of plasma. Of more general interest is the ability of the assay to cross react with plasma and pituitary extracts of non-gallinaceous birds. It can measure LH in the Tree Sparrow (*Passer montanus*), the Brambling, the White-crowned Sparrow, the Pekin duck, the Green-winged Teal (*Anas crecca*), and the European Starling, the domestic pigeon, the Herring Gull (*Larus argentatus*) and the canary. If it can be validated for these species, the assay should prove of wider value to avian endocrinology. Even with its special problems, the immunoassay appears superior to any available bioassay. The next few years will no doubt see assays developed for FSH and for the other tropic hormones of the adenohypophysis.

Chemistry

As yet, the avian gonadotropins have only been isolated from the domestic fowl, but it is encouraging that in this species they closely resemble mammalian LH and FSH in general chemistry and can therefore be purified by well-established methods. The following account summarizes the results of Stockell Hartree and Cunningham (1969) and of Scanes and Follett (1972).

The glycoprotein hormones (FSH, LH, and TSH) were extracted from broiler pituitaries (5-20,000) with a solution of ammonium acetate (6%) in 40% ethanol (pH 5.1). This is an efficient extraction procedure (>90%) which purifies the gonadotropins thirty- to sixty-fold. Growth hormone, prolactin, and ACTH remain in the insoluble pituitary residues. Further separation was effected on a column of carboxymethylcellulose. In dilute ammonium acetate solution (4 mM, pH 5.5) both LH and TSH are largely adsorbed into the cellulose network and form ionic bonds with the carboxylate groups. FSH remains unadsorbed and may be eluted directly. When the concentration of ammonium acetate is raised stepwise to 1 M (pH 5.5), LH and TSH come off together. This method is far from perfect and leaves some cross contamination of the gonadotropins. Subsequently, FSH was purified further by chromatography on columns of calcium phosphate, DEAE-cellulose, and Sephadex G-100. The material so produced has little LH and TSH activity.

A major problem has been the separation of LH from TSH, and as yet this has only been achieved by chromatography on DEAE-cellulose according to the method of Bates et al. (1968). The two hormones are very similar in their chemical properties, but under the right conditions, TSH can be bound preferentially to the column and a relatively TSH-free LH eluted. The LH fraction required further purification on columns of Amberlite IRC-50 and Sephadex G-100. Information on the amino acid composition of the LH is very similar to that of the mammalian LH's.

Physiology

The exact functions of avian FSH and LH have yet to be established, since the purified avian hormones have not been tested adequately in the hypophysectomized bird. However, experiments using mammalian gonadotropins suggest that in the male, FSH is responsible for growth of the seminiferous tubules, for the meiotic division of primary spermatocytes leading to secondary spermatocytes, and possibly for spermatogonial division. LH is generally regarded as acting primarily on the Leydig cells to stimulate production of androgen. For the attainment of full spermatogenesis, both hormones appear necessary; the reasons for this are unclear but may involve an intratesticular role for androgens in accelerating postspermatocyte development. In the female, the gonadotropins act to cause follicular growth as well as steroidogenesis, but the relative roles of FSH and LH within these processes have not been determined. As in the mammals, LH is regarded as the hormone that induces ovulation.

Preliminary experiments indicate that chicken LH, administered to quail, elicits a considerable testicular weight increase and causes differentiation of the Leydig cells.

The pituitary gonadotropin content is dependent on the state of sexual maturity of the bird, and in seasonally breeding species, the level undergoes a cycle that reaches its maximum on the breeding grounds. Photorefractoriness leads to a precipitate decline in pituitary content. The stimulation by long daily photoperiods of gonadal growth also leads to an increase in pituitary gonadotropins. Returning Japanese Quail to short daily photoperiods causes a decline in the pituitary content. Growth and the onset of sexual maturity in the chicken are likewise accompanied by an increase in pituitary gonadotropins. In all these situations a high pituitary level is associated with an elevated level of gonadotropin secretion as assessed by gonadal growth.

These findings are generally confirmed when the circulating LH level in Japanese Quail is measured by radioimmunoassay. Two points are perhaps worth emphasizing: (1) There is detectable LH in unstimulated young quail showing no testicular growth. Similarly, LH has been detected at low levels in the plasma of photorefractory White-crowned Sparrows; these observations suggest the hypothalamo-hypophyseal system may not be as inactive under these conditions as has often been assumed. (2) The plasma LH level is not correlated with the rate of testicular growth. Indeed, it remains relatively constant throughout the period of growth, although there is great individual variation. In the absence of data on the turnover of LH at different stages of sexual maturation, one cannot know whether the secretion rate remains stable or not over a month of photostimulation. There is no theoretical reason why the secretion rate needs to alter, but information on this point would be most valuable in understanding the dynamics of testicular growth and function in birds.

Under certain circumstances (rarely of natural occurrence) it is possible to separate gonadotropin synthesis in the pituitary from its release. The net result is that the pituitary level can be greatly elevated while testicular growth is either absent or at a low level.

Diurnal changes in pituitary gonadotropins have been measured in the Japanese Quail, and in the White-crowned Sparrow. Many of the experiments attempted to determine when the gonadotropic hormones are secreted under a stimulatory light schedule. This information could be important in establishing how the underlying circadian basis of avian photoperiodism is expressed endocrinologically. Unfortunately, the available results could support a variety of theories, and indeed, it seems unanswered yet whether there is a rhythm, or whether the changes may only reflect random fluctuations. The data are also subject to other problems, notably that they combine measurements of both LH and FSH, and that the "rhythms" might reflect storage rather than release of the hormone. Two further complications have arisen recently; the circulating LH level in a Japanese Quail represents only 0.5-1.5% of the stored pituitary content implying that quite large changes in plasma LH could occur with relatively minor alterations in the pituitary content. Second, Katongole et al. (1971) have shown in the bull that the plasma LH concentration cycles rapidly with five to ten peaks during a 24-hour period. The changes seem unrelated to daylight, feeding, or sleep. If this is a general phenomenon, it will make the detection of possible longer duration secretory rhythms a most difficult task. Daily rhythms in secretion must be distinguished from the type of data presented for the plasma LH levels in Greenfinches and House Sparrows. These indicate in the Greenfinch

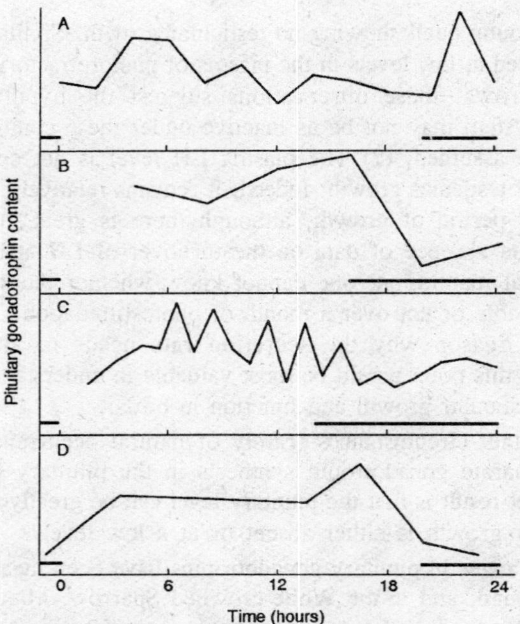

Fig. 1.3. Rhythms in pituitary gonadotrophin content of (A) Zonotrichia, leucophrys gambelii; (B), (C) and (D), Japanese Quail. The length of the dark period is shown by the black bars.

that LH tends to be secreted more readily than FSH under an asymmetric skeleton photoperiod of 6L: 0.5D: 1L: 16.5D, while the converse is true of the skeleton 6L:9D:1L:8D. The interpretation placed on these data is that LH and FSH may be secreted separately in time. The House Sparrow does not show the differential release.

As mentioned previously, castration, when accompanied by photostimulation, leads to pituitary hypertrophy and to an increase in the gonadotropin content. It also causes a fivefold increase in plasma LH in the Japanese Quail. If castrates are held on short daily photoperiods, these changes are not apparent. Treatment of intact or photostimulated castrates with exogenous sex steroids causes a rapid fall in pituitary and plasma LH.

Virtually all studies of the female have concentrated on the egg-laying cycle of the domestic fowl, and estimates of pituitary and plasma LH and FSH have attempted to establish the time at which these hormones are released in relation to ovulation. Neuroendocrine experiments have suggested that the release of LH for ovulation occurs

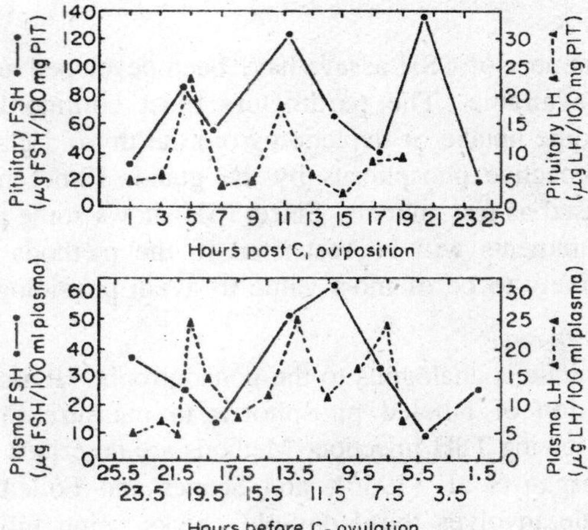

Fig. 1.4. FSH and LH levels in the pituitary and plasma of laying hens with respect to the times of oviposition and ovulation.

6-8 hours before ovulation itself, although some results of van Tienhoven et al. (1954) also suggest a burst of hormone release 14 hours prior to ovulation. In addition, one must expect that hormone secretion is associated with the maturation of follicles, a process that probably involves both hormones. The gonadotropin assays present a complicated picture with two to three peaks of both pituitary and plasma LH and FSH. Some difficulties of interpretation exist but the peak of plasma LH 8 hours prior to ovulation is assumed to cause that ovulation. The earlier peaks at 14 and 20 hours before ovulation are not fully understood. Imai and Nalbandov (1971) interpret their data to suggest a minor release of FSH at about the time of ovulation that may be essential for rapid growth of the next follicle, and a major release at about 14-11 hours prior to ovulation. The latter may be necessary for maturation and ovulation of the next follicle. Cunningham and Furr (1972) have recently shown, using the LH radioimmunoassay, that only a single peak of LH occurs in the plasma, some 8 hours prior to ovulation.

Thyrotropin (TSH)

The dependence of the avian thyroid gland on the adenohypophysis together with its responsiveness to mammalian TSH, makes it certain that an avian TSH exists. However, relatively little is known of the hormone; rather surprisingly so, because the young chick is used extensively for TSH bioassay. What evidence there is suggests that TSH has a physiological action similar to that in mammals.

Assay

A great number of TSH assays have been developed and extensive reviews are available. The parameters most commonly used are radioactive iodine uptake or depletion from the thyroid gland, and the uptake of radioactive phosphorus by the gland. Either mammals or birds can be used as test animals. Since TSH shows some phylogenetic specificity, comments will be restricted to the methods using birds and that are likely to be of most value to avian physiologists.

^{32}P-Uptake assay

This technique is analogous to the gonadotropin-^{32}P assay and uses the incorporation of labeled phosphorus to measure the thyroidal hyperplasia following TSH injection. Methods are described in Lamberg (1955), Greenspan et al. (1956), and Scanes and Follett (1972). A typical schedule involves the 1-day-old chicks being killed 7 hours after TSH injection (subcutaneously) and 1 hour after a dose of sodium [^{32}P] phosphate. The thyroid glands are removed, weighed, and dried prior to counting. The method is sensitive to 2.5 mU of ovine TSH (NIH-TSH-S6). With a (2+2) assay design and six chicks at each dose level, the mean index of precision from sixteen assays was 0.192 ± 0.017 (SEM). Chicken and quail pituitary extracts give dose-response curves parallel to that of the mammalian TSH standard.

^{131}I-Depletion assay

An advantage of this method is that the thyroid glands need not be dissected. The best method uses 1-day-old chicks from hens maintained on a low-iodine diet. The chicks are injected (subcutaneously) with 1-3μCi ^{131}I, and the uptake is measured the following day with a probe counter. Those showing an uptake greater than 15% are then injected with TSH together with 8 μg of thyroxine and 500 μg of a goitrogen such as propylthiouracil to prevent recycling of any liberated iodide. One day later, the depletion is measured relative to the ^{131}I present 24 hours earlier. The injections of TSH, thyroxine, and goitrogen are then repeated, and depletion is measured after another 24 hours. Since each chick is its own control, the errors are much reduced and an index of precision of about 0.2 may be expected. The dose range is generally between 0.5 and 4 mU TSH.

^{131}I-Uptake assay

Five injections of TSH are given at 12 hourly intervals to young cockerel chicks. This is followed by an injection of ^{131}I, the chicks being killed 5 hours later.

Radioimmunoassays for mammalian TSH are now available but not yet for avian TSH. Antisera raised against impure chicken TSH fractions can block the biological activity of chicken TSH when tested in the ^{32}P-uptake assay, but highly purified TSH suitable for radioiodination has yet to be prepared.

Identification

Pituitaries from a few avian species have been shown to contain TSH activity, viz., the chicken, turkey, pigeon, duck, canary, and the Japanese Quail. Circulating TSH has also been detected in the pigeon. Unfortunately, little is known either of the relative activity of avian TSH in the various bioassays or of the pituitary and plasma levels after physiological experimentation. Information is badly needed and should not prove too difficult to obtain with the assays now available. Chicken TSH has been partially purified. Chemically, it is similar to mammalian TSH, being a glycoprotein of large molecular weight.

Prolactin

Assay

Although a number of prolactin assays have been developed the most widely used continues to be that utilizing the crop-sac response in pigeons. This response to prolactin, which was originally demonstrated by Riddle et al. (1932, 1933), is remarkable; it apparently occurs only in birds of the order Columbiformes and represents a physiological adaptation akin to that of mammalian lactation. Prolactin stimulates a proliferation of the entire mucosal epithelium of the crop-sac. Eventually the innermost layers of epithelial cells slough off into the crop-sac lumen and form a "milk." Mixed with food, this is regurgitated and fed to the young.

Many attempts have been made to quantify the response as an assay procedure. These have usually varied in the route and schedule of the hormone injections and in the methods used to determine the degree of crop-sac stimulation. In the systemic method of assay, prolactin is injected once daily for 4 or 7 days into adult pigeons and the crops weighed the day after the last injection. The technique is rather insensitive, the minimally effective dose being about 500 mU. Sensitivity is much increased (100- to 200-fold) if the hormone is applied locally to the crop-sac by an intradermal injection, and since the crop-sac on each side can be injected separately, each pigeon yields two observations. The drawback to this method has been to quantify the local responses, and most techniques have assessed crop-sac development on a subjective scale. This has been improved upon by Nicoll (1967), who devised an apparatus to stretch the pigeon hemicrops by a uniform amount, thereby allowing comparable disks of tissue to be cut from each crop. The disks are dried and weighed. Responses are measured as a percentage weight increase of the hormone-injected hemicrop over that of the saline-injected side. With this quantitative measure, Nicoll obtained a satisfactory log dose-response relationship and an improved index of precision. Ben-David (1967) has quantified the crop-sac assay by measuring tritiated thymidine incorporation after prolactin treatment.

A further point worth considering has recently emerged. There is a pronounced daily rhythm in the crop-sac response to prolactin, injections late in the daily photoperiod being two- to fivefold more effective than at earlier times in the day. This suggests that a strict regimen should be employed when assays are to be carried out routinely, and also that certain schedules give greater sensitivity.

Even with these improvements, however, the crop-sac response leaves much to be desired. The confidence limits on the estimate of potency are invariably rather wide, and the assay is too insensitive to measure routinely the circulating prolactin concentration. Two other approaches have therefore been explored. The first utilizes the observation that prolactin is readily separable from the other adenohypophyseal hormones by polyacrylamide gel electrophoresis. By staining the prolactin band with aniline blue-black and measuring its density, it has been possible to develop a quantitative assay. The method is ten times more precise than the crop-sac assay, but much less sensitive. Even so, it can be used to detect the prolactin in single pigeon glands. Radioimmunoassays have now been developed in a number of mammalian species to measure prolactin. All are far more sensitive than any bioassay but are not generally applicable because of species specificity. Duck pituitary extracts do not cross-react in an ovine prolactin radioimmunoassay, while semipurified chicken prolactin is ineffective in the equivalent rat assay. It appears that a homologous radioimmunoassay will be needed to measure avian prolactin.

Physiology

The roles of prolactin are many, and the hormone has been implicated in a wide variety of processes, virtually all of which are associated directly or indirectly with reproduction. In many cases, prolactin does not act alone but requires the presence of other endocrine secretions in order to produce a complete response. Such synergistic effects, which are characteristic of prolactin action in other vertebrates, are well seen in the development of the brood patch, fat deposition and the induction of migratory restlessness. A detailed account of the physiology of the hormone cannot be included in this chapter, and comment will be restricted to those reports in which pituitary prolactin content has been measured and related to some event in the organism.

Many reports suggest that prolactin plays a part in the onset of broodiness, acting both to induce behavioural changes and, with the sex steroids, to cause hyperplasia of the brood patch. In support of this, the level of pituitary prolactin is 2.5-3 times greater in incubating than in nonincubating domestic hens. Similarly, the level in Ring-necked Pheasant hens (*Phasianus colchicus*) rises at the onset of incubation to reach its greatest value 8-12 days later. However, the prolactin content declines rapidly during the latter stages of incubation and the early post-hatching period, the time at which birds display their most intense breeding activity. This implies that assays on pituitary levels may not truly reflect the periods of highest secretion. Jones (1969) has compared pituitary prolactin content with the degree of incubation patch development in the California Quail (*Lophortyx californicus*). In wild females, the level is much the same for birds taken out of the breeding season and during the stages of ovarian growth, egg laying, and early incubation. The content increases to reach its highest level while the females show broodiness. The phase of incubation patch development precedes this rise and occurs at the onset of incubation, and by the brooding stage the patch is regressing rapidly. These results suggest that prolactin secretion is greatest during early incubation and that this keeps the hormone content at a low level. As the birds finish incubation, the demand for prolactin for growth of the brood patch diminishes, and this may account for the rise in the pituitary level. Caged egg-laying females showing no brood patch development and incubation behaviour have the highest prolactin levels of all. This may be an example in which prolactin synthesis occurs in the absence of significant release. Testicular growth is accompanied by a rise in prolactin content. In those males not forming an incubation patch, the prolactin level declined during testicular regression, but in those few that did form a patch, the content remained high. Here, then, is an example of a high level of prolactin secretion being correlated with a high pituitary content.

In the three species of the subfamily Phalaropodinae, only the male forms a brood patch and incubates the eggs. Again, prolactin is implicated (together with androgens); Hohn and Cheng (1965), using

Wilson's Phalarope (*Phalaropus tricolor*), found more prolactin in the pituitaries of males than of females. An increase in pituitary prolactin content has also been reported in incubating California Gulls (*Larus californicus*) and turkeys.

Pituitary prolactin cells are activated if birds are photostimulated, and this is accompanied by an increase in the hormone content of the gland. In the male duck, the prolactin content shows a pronounced annual cycle that very much parallels the cycle in testicular growth, the two parameters reaching their maximum levels simultaneously. Exposure of ducks to continuous light in the autumn likewise increases the prolactin level and induces testicular growth. Two species of quail behave in a similar manner to the duck. Some results of Gourdji (1970) suggest that the increase in prolactin content is a direct result of the increased day length and may be modified by changes in sex hormone secretion. Thus, castration of male ducks held on short days, or their treatment with exogenous testosterone, has no pronounced effects on pituitary prolactin content. Under conditions of continuous light, some effects were noted, with castration leading to a lower pituitary level. It is assumed that the high prolactin level found in these photostimulated birds reflects a high secretion rate, but little is known of the physiology of the hormone in these two species. Not surprisingly, both fall into that group of birds in which prolactin does not suppress gonadotropin secretion.

A number of papers from Meier's laboratory have given a prominent role to prolactin in the regulation of fattening and of migratory restlessness in passerine species. Most of the evidence derives from treatment with exogenous prolactin, and the fascinating discovery has been made that the time of administration of the hormone crucially affects the magnitude of the response. This could imply that under natural conditions the release of prolactin is more marked at one time of the day than another. Some evidence for this has been found in the White-throated Sparrow (*Zonotrichia albicollis*). During the period of vernal migration, the pituitary prolactin content shows a diurnal rhythm with a peak occurring 6 hours after dawn. Meier et al. (1969) interpret these data as indicating a release of prolactin during the afternoon. This is the same time of the day when exogenous prolactin is most effective in inducing fat deposition.

Adrenocorticotropin (ACTH)

Indirect evidence for an avian ACTH has existed for many years, but surprisingly, it is only recently that its presence in the adenohypophysis has been established. The presence of ACTH in anterior and posterior pituitaries of the Pekin duck has been shown using the adrenal ascorbic acid depletion in the hypophysectomized rat. Pituitary extracts from the chicken, duck, and pigeon have all proved capable of stimulating corticosteroid production *in vitro* by adrenal tissues of both mammals and birds. An intravenous injection into chickens of a crude chicken pituitary extract also raises the adrenal venous plasma concentration of corticosterone within 30 minutes. Rather more unequivocal evidence for chicken ACTH has come recently from Salem et al. (1970a,b). These authors used a well established ACTH bioassay based on adrenal ascorbic acid depletion (AAAD) in the hypophysectomized rat. In this system, chicken pituitary extracts gave a dose-response curve parallel with the purified mammalian standard, and the ACTH activity of a laying hen was estimated at 401.7 (270.0-597.7) mU per milligram of frozen tissue. This is somewhat higher than that reported by De Roos and De Roos (1964). The assay sensitivity is such that a complete (2 + 2) assay should be possible with 0.01 of a chicken pituitary, suggesting the AAAD method might be adequate for assays in single pituitaries of small birds.

Avian ACTH, like the mammalian hormone, is labile to boiling and to rapid adjustments in pH. This is in contrast to the ACTH-like activity that resides in the avian hypothalamus. The functional role of the hypothalamic ACTH-like substance is still a matter of speculation.

Melanotropin (MSH)

Investigations on the melanotropin content of the bird pituitary have been directed toward the following: (1) appearance of this hormone during the development of the chick; (2) distribution of the hormone within the pars distalis of the chick and the duck; (3) correlation between the pituitary level of MSH and the abundance of κ cells. For these purposes, quantitative data have been obtained using different bioassays—the hypophysectomized lizard *Anolis*, the hypophysectomized male frog, and Burger's method *in vitro* with the skin of *Anolis carolinensis*. The specificity of these bioassays is not perfect, since they display some sensitivity to ACTH.

The MSH activity is probably the first to appear during the embryological life of the chick—on the 5th day. Its concentration in the pituitary increases very rapidly during the second half of the incubation but remains unchanged after hatching. Within the gland, the hormone is more concentrated in the cephalic lobe of the chick and the duck (thirty times more). In some species of duck and pigeon, there is a higher level of pituitary MSH in strains with coloured feathers as compared to strains with white feathers. This finding suggests that MSH could intervene in the control of feather pigmentation. To our knowledge, this is the only indication as to a possible physiological role of this hormone in birds. As of now, the chemistry of avian MSH is not known.

Growth Hormone (Somatotropin, STH)

Pituitary extracts of chicken and turkey possess some growth-promoting activity since when injected into immature hypophysectomized rats, they cause widening of the tibial epiphyses, and increases in body weight and nitrogen retention. In this assay, a rat pituitary contains about eight times as much activity as that of a chicken. This might reflect a low level of growth hormone in the chicken, but may equally well reflect a species difference in its structure and activity. A comparable phenomenon occurs with fish growth hormone, which is highly active in hypophysectomized fish but inactive in the rat. Chicken pituitary extracts do increase growth rate, nitrogen retention, and bone growth in hypophysectomized chicks, but the mammalian growth hormones of bovine and porcine origin have no effect even at large dose levels. Again, this may reflect a species specificity, since mammalian growth hormones alter plasma protein levels in the Budgerigar (*Melopsittacus undulatus*), while pituitary tumors from Budgerigars cause growth in rats. Immunologically, antibovine growth hormone shows no cross reaction with chicken pituitary extracts.

2

NEUROHORMONES

External and internal environmental changes may stimulate the receptors of animals. Some of the information generated at the receptors in the form of nerve impulses may be transferred ultimately to neurosecretory cells in the hypothalamus. The neurosecretory cells transform the afferent neural signals into neurohormonal information through neurohypophyseal hormones and adenohypophyseal hormone-releasing factors. The former are released into the systemic circulation via the neural lobe and act mainly on the kidney, blood vessels, and uterus. This neurosecretory system can be demonstrated by means of Gomori's aldehyde fuchsin or chrome alum hematoxylin staining methods. The sites of production of the neurosecretory material are well known. The releasing factors are accumulated in the median eminence and pass into the adenohypophysis through the portal vessels. The neurosecretory system producing the releasing factors is not, or at least largely not, stained with Gomori's staining methods. Therefore, the location of the nuclei that produce the releasing factors and the pathways involved have not yet been determined. In addition to these two neurosecretory systems, the ependymal cells of the median eminence seem to be involved in the transport of information from the third ventricle to the adenohypophysis. The main motive of recent research in neuroendocrinology has been to discover the series of events in the processes mentioned above and to clarify the mechanisms whereby the processes are integrated.

Investigations in this field are most advanced in mammals, and much less is known in birds. In order to understand neuroendocrine mechanisms in general, it is necessary to know basic patterns common to animals of all vertebrate classes and, further, to recognize characteristic features that are specific to animals of a certain class. In birds, several characteristic features have already been demonstrated: (1) nerve cells in the hypothalamus are generally diffusely distributed; (2) the median eminence is morphologically distinct; (3) the adenohypophysis is completely separated from the brain; (4) portal vessels generally lie external to the median eminence and the adenohypophysis; and (5) gonadal growth is readily regulated by photoperiodic manipulation in some species. Except for (1), these features are advantageous for studying neuroendocrine mechanisms in birds. Some investigators have considered these advantages and engaged in avian neuroendocrinology, although the number is much smaller than those that investigate mammalian species. Excellent reviews of the avian studies have been published by Earner and Oksche (1962) and Earner et al. (1967). Therefore, papers published before 1962 are not cited in this chapter unless they are specifically necessary. Further, the pineal and subcommissural organs are not mentioned here, since they are described in other chapters.

Anatomy of the Avian Neurosecretory System

Light Microscopy of the Neurosecretory System

Neurosecretory cells and their axonal pathways

Gomori-positive neurosecretory system. The hypothalamic neurosecretory nucleus of fishes and amphibians consists of one cell group, the preoptic nucleus. In fishes, the neurosecretory cells send their axons mostly to the posterior part of the neurohypophysis, which is equivalent to the pars nervosa of higher vertebrates, but a few axons terminate in the anterior part, which corresponds to the median eminence. In amphibians, the neurosecretory cells send their axons to the median eminence and the pars nervosa. In fishes and amphibians, these axons form the preopticohypophyseal tract. In reptiles and birds, the hypothalamic neurosecretory nuclei consist of the supraoptic nucleus and the paraventricular nucleus. The neurosecretory neurons of both nuclei send axons to the median eminence and the pars nervosa. The axonal tract, which is clearly seen in the median eminence, leading to the pars nervosa is designated as the supraopticohypophyseal tract. It should be noted here that the median eminence of reptiles and birds is so well developed that it is excellent material for morphological investigations.

Several investigators have demonstrated that both the supraoptic and paraventricular nuclei contain several clusters of neurosecretory

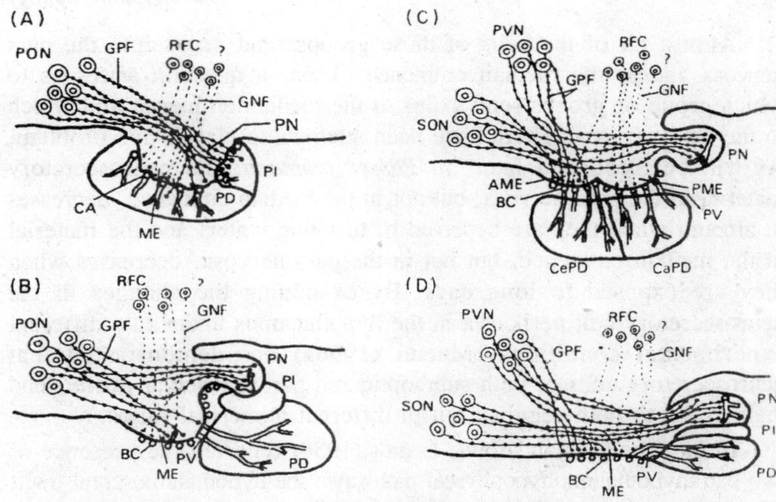

Fig. 2.1. Diagram of sagittal section of the median eminence-hypophyseal region, (A) teleost; (B) anurans; (C) reptiles and birds; (D) mammals. AME, anterior median eminence; BC, capillaries of primary plexus; CA, capillaries draining into pars distalis; CaPD, caudal pars distalis; CePD, cephalic pars distalis; GNF, Gomori-negative fiber; GPF, Gomori-positive fiber; ME, median eminence; PD, pars distalis; PI, pars intermedia; PME, posterior median eminence; PN, pars nervosa; PON, preoptic nucleus; PV, portal vessel; PVN, paraventricular nucleus; RFC, cells producing releasing or inhibiting factors; SON, supraoptic nucleus.

cells. In the Budgerigar (*Melopsittacus undulatus*), for instance, three neurosecretory cell groups are recognized in the supraoptic nucleus. Matsui (1966a) divided the neurosecretory cells into ten groups in the Tree Sparrow (*Passer montanus saturatus*). In *Zosterops japonica*, Arai (1963) and Uemura and Kobayashi (1963) divided the neurosecretory neurons into several groups. The number of neurosecretory cells of each group in *Zosterops* has been estimated by Arai (1963). Rossbach (1966) identified nine groups of neurosecretory cells in the hypothalamus of the European Blackbird (*Turdus merula*): nucleus entopeduncularis anterior, nucleus entopeduncularis medialis, nucleus entopeduncularis posterior, nucleus entopeduncularis ventralis, nucleus lateralis externus hypothalami, nucleus magnocellularis interstitialis dorsalis, nucleus paraventricularis dorsalis, nucleus paraventricularis ventralis, and nucleus supraopticus medialis lateralis. Thus it is typical for the neurosecretory cells to occur in several groups or clusters. However, functional differences among these groups of neurosecretory cells are not known.

Almost all of the cells of those groups send axons into the pars nervosa and/or the median eminence. Then, a question arises as to which group or groups send axons to the median eminence and which to the pars nervosa. There have been attempts to clarify this problem. As will be mentioned later, in *Passer montanus* the neurosecretory material in the pars nervosa, but not in the median eminence, decreases in amount when they are deprived of drinking water, and the material in the median eminence, but not in the pars nervosa, decreases when they are exposed to long days. By examining the changes in the neurosecretory cell perikarya in the hypothalamus under two different experimental conditions, Matsui (1966a) has demonstrated that neurosecretory cells of both supraoptic and paraventricular nuclei send their axons to both regions, but in different numerical ratios.

In the Rhode Island fowl, Legait (1959) reported the presence of two extrahypothalamohypophyseal pathways: the hypothalamoseptal tract, which ends in a subseptal organ (subfornical organ), and the hypothalamohabenular tract, which connects the hypothalamic and epithalamic regions of the brain. The functional significance of these pathways is not known.

Gomori-negative neurosecretory system. It has recently been demonstrated in mammals that the brain produces hypophysiotropic neurohormones that induce or inhibit hormone release from the adenohypophysis. However, specific sites in the brain for the production of these releasing and inhibiting factors are not known. Judging from pituitary cytology, the avian adenohypophysis seems to be similar to that of the mammals and apparently produces six hormones. Therefore, at least six hypophysiotropic neurohormones must be produced in the brain. Actually, adrenocorticotropic hormone-, gonadotropin-, prolactin-, and growth hormone-releasing factors have been demonstrated in birds. However, the nuclei that produce these neurohormones have not yet been determined. Detailed physiology of the releasing factors in birds will be mentioned later.

Median eminence

A definition of the median eminence has recently been presented by the senior author. According to him, the median eminence is exteriorly the portion covered by the capillaries of the primary plexus of the hypophyseal portal veins, and interiorly the basal portion of the hypothalamus occupied by the processes of the secretory ependymal cells. In this portion there are some perikarya of the infundibular nucleus. However, these neurons are excluded from the definition of

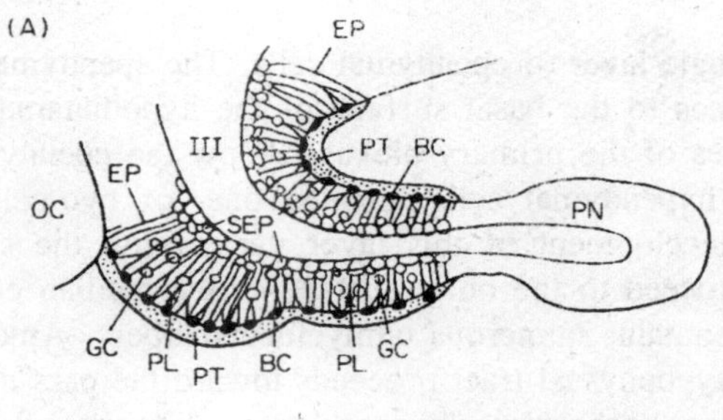

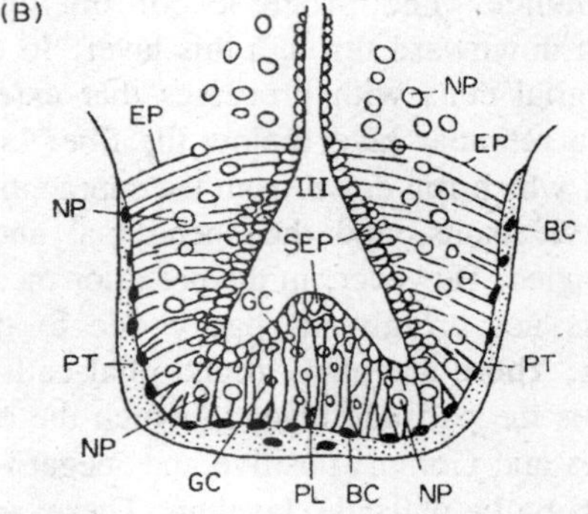

Fig. 2.2. Diagram of the median eminence of the pigeon, (A) sagittal section, (B) cross section. BC, capillaries of primary plexus; GC, glial cells; EP, ependymal processes; NP, perikaryon of neuron; OC, optic chiasma; PL, palisade layer; PN, pars nervosa; PT, pars tuberalis; SEP, secretory ependymal cells; III, third ventricle.

the median eminence in order to generalize its definition. For instance, die anterior portion of the teleost neurohypophysis, the equivalent of the median eminence of higher vertebrates, does not contain any neurons. Usually, the median eminence begins anteriorly immediately caudal to the optic chiasm and ends at the beginning of the infundibular stem, on the surface of which the capillaries of the primary plexus no longer occur. The avian median eminence is clearly divided into anterior and posterior divisions. The anterior division contains abundant aldehyde fuchsin-positive neurosecretory material, whereas the posterior division rarely contains such material.

In the median eminence, the following layers can be recognized: (1) ependymal layer, (2) hypendymal layer, (3) fiber layer, (4) reticular layer, and (5) palisade layer. On the basal surface of the palisade layer lie the capillaries of the primary plexus. Further, the pars tuberalis covers the network of the primary plexus. The ependymal layer is a single layer of ependymal cells. The ependymal cells extend their processes to the basal surface of the hypothalamus and also to the capillaries of the primary plexus. Below the ependymal layer are located the hypendymal cells forming one- or two-cell layers. The degree of development of this layer varies with the species. Their processes proceed to the outer surface of the median eminence. The fiber layer contains numerous unmyelinated fibers. Among them, the supraopticohypophyseal tract proceeds toward the pars nervosa. From this tract, many neurosecretory axons proceed down toward the surface of the median eminence. The processes of the ependymal and hypendymal cells run downward through this layer. In the fiber layer there are elongated glial cells with processes that extend in various directions. There is a reticular layer below the fiber layer; here, the neurosecretory axons, which run down from the supraopticohypophyseal tract, Gomori-negative axons, and the ependymal and hypendymal processes are intermingled. However, in the posterior median eminence, the reticular layer is not prominent, partly due to the paucity of neurosecretory axons. There are many round glial cells in this layer. The outermost layer is the palisade layer in which the ependymal and hypendymal processes and Gomori-positive and -negative nerve fibers are arranged in a radially palisade fashion. There are many glial cells, and the neurosecretory material is accumulated around them, just as in the neural lobe.

Neural lobe

Wingstrand (1951) grouped the avian neural lobe into four categories according to the thickness of the diverticular wall and the size of the lumen. The primitive neural lobe has many thin-walled diverticula and a large lumen (*Gallus*, *Phasianus*). The most complex is that with thick-walled diverticula and with a very restricted lumen (Anseriformes, *Larus*). The thin diverticular wall of the neural lobe consists of an ependymal layer facing the lumen, a fiber layer, and a "glandular zone" (external layer) that contains aldehyde fuchsin (AF)-positive neurosecretory axon endings. The connective tissue and the vessels are on the surface of the external layer. The glial cells (pituicytes) are distributed in the fiber and external layers. However, the neural lobe with the thick-walled diverticula contains more neural and glial elements and is very compact in appearance. In all cases, the neural lobe contains numerous terminals of neurosecretory axons and, therefore, has a strong affinity for aldehyde fuchsin or chrome alum hematoxylin.

The neurosecretory material is dense around the glial cells, especially around those cells that are near blood vessels. This suggests some functional relationship between the glial cells and the neurosecretory axons. Immediately adjacent to the blood vessels there is a clear zone in which neurosecretory material is almost absent. Electron microscopy has revealed that this zone is composed of the terminals of the pituicyte and ependymal processes as mentioned later. In other words, the ependymal and pituicyte processes usually intervene between the neurosecretory axon endings and the capillaries in the neural lobe. These findings raise a question as to whether or not the ependymal and glial (pituicyte) processes are involved in the release of neurosecretory substance from the neurosecretory endings into the capillaries.

It should be noted here that destruction of the neurohemal organs of the stainable hypothalamohypophyseal neurosecretory system in the neural lobe and/or palisade layer of the anterior median eminence of the White-crowned Sparrow (*Zonotrichia leucophrys gambelii*) results in the formation of large aggregations of stainable neurosecretory material surrounding the hypertrophied capillaries. In the Japanese Quail, this appears not to occur and there is no indication of repair. It seems that the degree of regeneration of the neurosecretory system depends upon the species involved.

Fiber connections in the neurosecretory system

Wingstrand (1951) described four efferent hypothalamic tracts terminating in the neurohypophysis of the pigeon: (1) tractus (tr.) hypophyseus anterior, (2) tr. supraopticohypophyseus, (3) tr. tuberohypophyseus, and (4) tr. hypophyseus posterior. The tr. supraopticohypophyseus is formed by neurosecretory fibers that originate from the supraoptic and paraventricular nuclei. As already mentioned, some fibers of this tract proceed to the anterior median eminence, but most fibers terminate in the neural lobe. The tr. tuberohypophyseus is formed by the fibers of the infundibular nucleus. The tr. hypophyseus anterior seems to include all hypophyseal fibers behind the chiasma which do not belong to the tr, supraopticohypophyseus or the tr. tuberohypophyseus. However, it is possible that the tr. hypophyseus anterior contains a part of a bundle in the supraopticohypophyseal tract, which is derived from the most rostral neurosecretory cells. The tr. hypophyseus posterior proceeds to the pars nervosa. This tract is distinct in *Anser*. However, the origin of the fibers of this tract is not known.

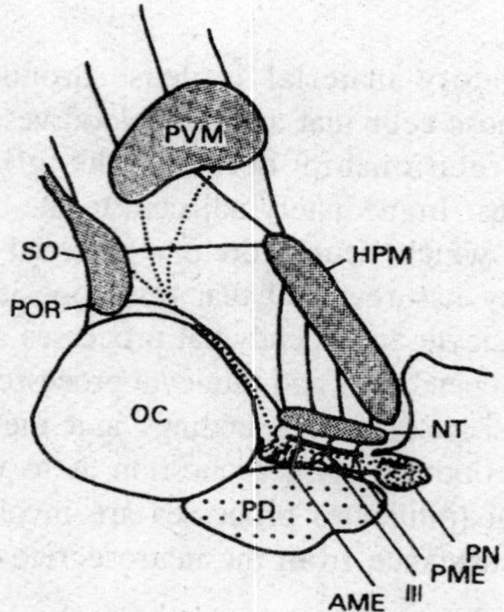

Fig. 2.3. A diagram of a sagittal section of the hypothalamus of the Japanese Quail showing the regions containing monoamines and the fiber connections between them. HPM, nucleus hypothalamicus posterior medialis; NT, nucleus tuberis; POR, preoptic recess; PVM, nucleus paraventricularis magnocellularis; SO, nucleus supraopticus.

Through the use of fluorescence microscopy, monoaminergic fibers terminating in the median eminence have recently been demonstrated. The median eminence receives the fluorescent fibers from the nucleus tuberis (NT) and possibly from the nucleus hypothalamicus posterior medialis (NHPM). These fibers form part of the tuberohypophyseal tract. The two nuclei (NT and NHPM) are connected by two tracts; one tract runs in the stratum cellulare internum close to the wall of the third ventricle and the other curves round through the lateral regions of the hypothalamus.

Monoaminergic fibers associated with the Gomori-positive axons proceed down behind the posterior border of the optic chiasma toward the median eminence in the Japanese Quail, the White-crowned Sparrow, and the House Sparrow (*Passer domesticus*). These fibers must form in part the tractus hypophyseus anterior of Wingstrand. This fiber tract can be traced to the supraoptic nucleus. Oehmke (1969) observed green-yellow fluorescence (indicating the presence of monoamines) in the infundibular and ventromedial nuclei of the Greenfinch (*Carduelis chloris*) and the Mallard (*Anas platyrhynchos*). He divided the infundibular nucleus into three regions. The basal layer showed stronger fluorescence than the middle and dorsal layers. Fluorescence was also observed in the subependymal zone of the median eminence of both species, as well as other birds. The fluorescence in the palisade layer shows considerable variation in the intensity depending on the species. The fluorescent fibers in the palisade layer are probably derived from the monoaminergic fibers in the subependymal layer, which originate from the nucleus tuberis and possibly from the nucleus hypothalamicus posterior medialis. In the reticular layer, a few fluorescent fibers run in a rostrocaudal direction; these can be traced to the neural lobe in the domestic fowl and the Japanese Quail.

The neurosecretory cells of the supraoptic and paraventricular nuclei are embedded in monoaminergic nerve fibers. Monoamine-oxidase activity is also strong around the neurosecretory cells of the Japanese Quail. These monoaminergic fibers have been found to form synaptic contacts with the neurosecretory cells. The origin of these fibers is still unknown, but it has been suggested that the fibers may be derived from those in the forebrain bundle as was observed in mammals. In the pigeon and the Japanese Quail, the fluorescent tract in the forebrain bundle continues anteriorly to the nucleus basalis and the forebrain.

There are monoaminergic fiber connections between the following regions, as mentioned above: nucleus (n.) tuberis, n. hypothalamicus posterior medialis, n. paraventricularis, n. supraopticus, preoptic recess, n. basalis, and median eminence. These different monoaminergic fiber connections in the avian hypothalamus may be a part of the integrative mechanism in the hypothalamic control of the adenohypophysis.

Cholinergic innervation in the median eminence and neural lobe remains obscure, but acetylcholinesterase has been shown in the neurosecretory cells, cells of the n. infundibularis, the palisade layer of the median eminence, and the pars nervosa.

Electron Microscopy of the Neurosecretory System

Since the nuclei that produce the hypophysiotropic neurohormones have not yet been identified, this description will be confined mostly to the Gomori-positive neurosecretory system.

Neurosecretory cells and their fibers

The neurosecretory cells of the supraoptic nucleus of the House Sparrow contain 2000-2500 Å granules. The granules are formed in the Golgi apparatus. The neural lobe contains 2000-2500 Å granules. Therefore, 2000-2500 Å granules must be carriers of the neurohypophyseal hormones. Priedkalns and Oksche (1969) found synaptic terminals of the monoamine fibers at the cell perikarya of the supraoptic nucleus and the infundibular nucleus of *Passer domesticus*.

In the neuropile and the perikarya of the nucleus infundibularis, in the tuberoinfundibular tract and in the reticular and palisade layers of the median eminence, 500-1000 Å granules are found in *Passer domesticus*. Since these granules disappeared after reserpine treatment, they seem to contain monoamines.

Anterior median eminence

There have been several investigations of the fine structure of the avian median eminence. Some ependymal cells have microvilli and bulbous protrusions on the apical surfaces in the White-crowned Sparrow, duck, and pigeon. The physiological meaning of this phenomenon will be mentioned later. In the subependymal layer, the hypendymal (subependymal) cells show a structure similar to the ependymal cells except for the apical protrusions. Among the processes of the ependymal cells in this layer there are sometimes fibers containing granules of 1000 Å. The presence of this type of granule coincides with the presence of monoamine fluorescence in the subependymal portion. In the fiber layer, there are many unmyelinated axons, each containing small (1000 Å), intermediate (1200-1500 Å), or large (1500- 2500 Å) electron-dense granules. In addition, in this layer, small axons without granules are numerous. The large granules and intermediate granules occur in the neural lobe. Therefore, these granules are doubtless carriers of the neurohypophyseal hormones. The small granules are probably carriers of monoamines. However, there is the possibility that these three kinds of granules contain releasing factors.

In the reticular layer, there arc axons containing the small, intermediate, or large granules. These three kinds of axons and processes of the ependymal, hypendymal, and glial cells are interwoven in this layer.

In the upper portion of the palisade layer, there are fibers or processes that are continuations from the reticular layer. They proceed perpendicularly to the ventral surface of the median eminence. The axons with or without the three kinds of granules in this layer frequently contain synaptic vesicle-like structures with diameters of 300-500 Å. The axons containing exclusively these vesicles may be cholinergic fibers.

The peripheral portion of the palisade layer of the anterior median eminence consists of numerous unmyelinated fibers with or without the granules, nerve endings, and processes of glial and ependymal cells. Most of the nerve fibers contain one of three kinds of granules and synaptic vesicle-like structures. However, the fibers containing large granules are far fewer than in the upper portion. In the White-crowned Sparrow, *Passer domesticus*, *Agelaius tricolor*, and *Zonotrichia atricapilla*, the feet of the ependymal processes are frequently interposed between the nerve endings and the perivascular space of the primary plexus. However, in the Budgerigar and the duck, both the nerve endings and the ependymal processes are in contact with the perivascular space. The biological significance of this difference is obscure.

The perivascular space of the capillaries covering the surface of the median eminence is very thick (about 1 μ), a condition generally true of other endocrine organs. The space may be a reservoir for neurohormones. Sometimes it protrudes into the parenchyma of the median eminence, thus increasing the contact area between the space and the parenchyma. There are parenchyma] and endothelial basement membranes in the space, and among them are fibroblasts and many collagen fibrils. The endothelial cells of the capillaries have many pinocytotic vesicles on the cell membrane facing the perivascular space. This suggests that neurohormones that are released into the perivascular space from the axons are absorbed by the endothelial cells and then released into the capillary lumen.

Posterior median eminence

The fine structure of the posterior median eminence is basically the same as that of the anterior median eminence. The principal differences are that (1) in the former there are few fibers that contain the large and intermediate granules in the reticular and the palisade layers; (2) synaptoid contacts between the axons containing granules of 1000 Å and the ependymal processes are frequent in the hypendymal and palisade layers in the posterior median eminence but seldom in the anterior median eminence; (3) in the posterior median eminence, the axon endings are more frequently in contact with the perivascular space than in the anterior median eminence; (4) the perivascular space of the posterior median eminence protrudes more digitations in complex form into the parenchyma than in the anterior median eminence.

Neural lobe

Duncan (1956) first examined the avian neural lobe with an electron microscope. In the neural lobe of the chicken he observed many neurosecretory endings containing many granules with diameters of 1000-2000 Å and pituicytes and their processes. In 1959, Legait and Legait observed granules of 1000-1800 Å in the pituicytes of the fowl neural lobe. Both investigators pointed out the intimate contact of the neurosecretory fibers with the pituicytes and suggested the involvement of the pituicyte in the release of the neurohypophyseal hormones from the endings. Since these investigations, electron microscopy has been done on the neural lobe of several species of birds—Budgerigar, White-crowned Sparrow, and pigeon. All the investigators showed the presence of two kinds of granules therein; one with diameters of about 1200 Å and the other about 1600 Å. Interposition of the feet of the pituicytes between the nerve endings and the pericapillary space is not always found. It depends on the species. The axons in the neural lobe contain the small vesicles of about 500 Å in addition to the granules. There are some axons containing only the small vesicles of about 500 Å, the granules being absent. These axons may be cholinergic fibers terminating in the neural lobe.

With fluorescent microscopy, it was found that in the reticular layer of the Japanese Quail monoaminergic fibers proceed to the neural lobe. Therefore, it is expected that the granules of about 1000 Å should be present in the avian neural lobe.

The perivascular space of the neural lobe is thick as in the median eminence. There are many pinocytotic vesicles on the surface of the endothelial cells facing the perivascular space. The vesicles may be involved in the transendothelial transport of the neurohypophyseal hormones.

Hypothalamohypophyseal Neurovascular Link

The details of the blood supply and drainage of the hypothalamohypophyseal region have been described by Wingstrand (1951) and Assenmacher (1952, 1953). More recently, Vitums et al. (1964) have described in detail the distribution of the blood vessels of the hypothalamohypophyseal system. Their most interesting observation is the distribution of the capillaries of the primary plexus on the surface of the median eminence. There are distinct anterior and posterior capillary plexuses corresponding to the anterior and posterior divisions of the median eminence. Anterior and posterior groups of portal vessels collect the blood from the anterior and posterior primary capillary plexuses, respectively. The anterior group of portal vessels is mainly distributed to the cephalic lobe of the pars distalis and the posterior group of the portal vessels supplies the caudal lobe of the pars distalis. The development of the portal blood system has been described in the White-crowned Sparrow by Vitums et al. (1966). Sharp and Follett (1969a) have also observed that the primary plexus of the median eminence of the Japanese Quail is divisible into anterior and posterior parts and that it shows a point-to-point distribution to the pars distalis. Dominic and Singh (1969) have also observed the same phenomena in fifteen species of birds. Since the cell types of the cephalic lobe of the pars distalis are not always the same as those of the caudal lobe, it is supposed that different adenohypophyseal hormone releasing factors may be delivered to the anterior and posterior median eminence.

Duvernoy et al. (1969) studied the organization of the primary plexus of several species of birds. According to them, the primary capillary bed is composed of superficial vessels and deep vessels. The deep vessels include capillary loops and a subependymal network, both of which anastomose with hypothalamic vessels. They argue that the deep vessels connect the primary plexus and the third ventricle. This idea is important because there is the possibility that the ventricular fluid contains adenohypophyseal hormone releasing factors in mammals.

Biologically Active Substances in the Neurohypophysis

As mentioned before, Gomori-positive neurosecretory material is stored in the median eminence and the pars nervosa in birds. The avian neurohypophyseal hormones in the neural lobe are arginine vasotocin and oxytocin. Since the function of the median eminence is different from the pars nervosa, the possibility was expected that the Gomori-positive neurosecretory material in the median eminence would contain different neurosecretory hormones from those of the pars nervosa. However, it was found by bioassay methods and chromatography that the median eminence contains the same neurohypophyseal hormones as those of the pars nervosa. This is true for the animals of other vertebrate classes.

The neurohypophyseal hormones appear to be present in the large and intermediate granules. It is not known which type of granule contains which neurohypophyseal hormone (arginine vasotocin or oxytocin). Moreover, there is the possibility that these granules also contain releasing factors. Bioassays and chemical analyses of substances in the granules isolated by ultracentrifugation may yield more precise information on the nature of the substances in the granules. Carrier proteins of the neurohypophyseal hormones have not yet been studied in birds, but they seem to contain cysteine. DL-Cysteine[^{35}S] injected into the third ventricle was taken up by the tissues containing Gomori-positive material.

The small granules of about 1000 Å may be carriers of monoamines, since the small granules in the neural lobe and the median eminence are similar in size and profiles to the granules carrying monoamines in the anterior hypothalamus of the rat. The distribution of the small granules is almost the same as that of monoamines and monoamine oxidase in the median eminence.

With respect to the small vesicles in the median eminence and pars nervosa, there is lack of agreement about their nature in the mammalian neurohypophysis. This point is also being discussed with reference to the avian neurohypophysis. Some investigators are inclined to believe that they are carriers of acetylcholine (ACh), whereas others argue that they might be derived from the fragmentation of large granules. Herlant (1967) is of the opinion that there are two types of small vesicles— (1) synaptic vesicles carrying ACh and (2) small vesicles derived from the budding of neurosecretory granules. Furthermore, he has suggested that the small vesicles produced by budding contain neurosecretory material and that they are finally expelled into the pericapillary space. Since acetylcholinesterase is present in the median eminence and neural lobe, ACh should also be present there, possibly being carried by the vesicles.

As will be mentioned below, the adenohypophysis secretes at least six hormones. Therefore, it is naturally expected that the hypothalamus produces at least six neurohormones (releasing or inhibiting factors) that regulate the adenohypophyseal hormone secretion. However, in the avian median eminence, only three kinds of granules were recognized. Therefore, it is possible that granules with the same diameter carry at least two different releasing factors. For instance, corticotropin releasing factor is present in granules larger than 1500 Å in diameter, which were isolated by ultracentrifugation from the horse median eminence. These granules also possess vasopressor activity. Thus, there are neurohypophyseal hormones, monoamines, acetylcholine, and releasing factors in the avian median eminence. The physiological significance of these substances, except for releasing factors, is not known at the present time.

Functional Aspects of the Gomori-Positive Neurosecretory System

Water Metabolism

The presence of an antidiuretic substance in the avian neurohypophysis has been known for a long time. The neurosecretory system becomes active following dehydration or osmotic stress in birds: the neurosecretory material decreases from the pars nervosa, but not from the median eminence; and the neurosecretory cells increase their nuclear diameters. Vasotocin has no effect on the filtration rate but only on the tubular reabsorption of water. It is interesting to note that the Budgerigar and Zebra Finch (*Poephila guttata castanotis*), which live in the hot, dry parts of Australia, respond weakly or not at all to water deprivation.

Gonadal Growth

Studies involving sectioning of the supraopticohypophyseal tract of the duck and lesions of the neurosecretory nuclei lead to the suggestion that the hypothalamic neurosecretory system, including both the supraoptic and paraventricular nuclei, is involved in the control of gonadotropic activity of the adenohypophysis. These important investigations carried on in Benoit's laboratory are now rather classic; they have stimulated investigations in many laboratories. In the fowl, Legait (1959) observed changes in activity of the neurosecretory cells during periods of sexual inactivity, egg-laying, incubation, and molting. Mikami (1960) reported that in the domestic fowl the neurosecretory cells change their activity after castration. Graber and Nalbandov (1965) found, however, that castration in the fowl has little effect on the neurosecretory system. In the passerine birds, Oksche et al. (1959) performed extensive studies on this problem using the White-crowned Sparrow. The neurosecretory system is relatively inactive during the midwinter period of sexual inactivity, and the neurosecretory material is accumulated in the perikarya and in the external layer of the median eminence. During the early summer period of sexual activity, the neurosecretory material decreases in amount in the perikarya and in the median eminence. During the refractory period, the neurosecretory material is stored extensively in both the cells and the median eminence. These changes in the neurosecretory system were confirmed under natural photoperiods in the same species. Rossbach (1966) has also observed seasonal changes in the hypothalamic neurosecretory system. Konishi and Kato (1967) found a similar rhythm of the neurosecretory activities of the cells and the median eminence in the Japanese Quail. The rhythm could be changed by artificial daily photoperiods. Konishi (1965, 1967) found that neurosecretory material is more abundant in the median eminence of Japanese Quail kept under short days than that of quail held in continuous light. He observed, however, that there was no correlation between the amount of neurosecretory material and the testicular development. Ishii et al. (1962) found that there was an increase of neurosecretory material in the median eminence of the duck in which the growth of the gonads was stimulated by long days. Hirano et al. (1962) found that the amount of neurosecretory material increased in the median eminence, but not in the pars nervosa, of *Zosterops japonica* subjected to long days. Thus, the results obtained by different investigators concerning the involvement of the neurosecretory system in gonadal growth are contradictory.

As mentioned before, the hypothalamic neurosecretory nuclei are topographically divisible into several cell groups. Uemura and Kobayashi (1963) divided the neurosecretory cells of *Zosterops* into seven groups: the supraoptic nucleus was subdivided into lateral and median groups; the paraventricular nucleus was subdivided into anterior, periventricular, and lateral groups. The sixth group was distributed in the peduncles, and the seventh group was located in the hilar region of the median eminence. When the birds were subjected to long days, the cells of the lateral group of the supraoptic nucleus and those of the anterior and periventricular groups were stimulated, but the other groups were not. Matsui (1966a) has also obtained similar results in *Passer montanus*. In *Zosterops*, estrogen administration activated the neurosecretory cells of the median group of the supraoptic nucleus of the long-day birds and nullified the activating effect of the long days on the neurosecretory cells of the lateral group of the supraoptic nucleus. Rossbach (1966) observed the parallelism of seasonal activity between the nucleus paraventricularis and the gonadal cycle, and between the nucleus magnocellularis interstitialis dorsalis (neurosecretory) and adrenal cortical tissues. Other neurosecretory nuclei did not show any correlation with the activities of other endocrine glands. Thus, there are functional differences among different neurosecretory cell groups, and some cell groups show activity changes during gonadal growth, whereas others do not. Since releasing factors for the adenohypophyseal hormones have been found, the changes in neurosecretory activity during gonadal growth may not be directly related to the release of adenohypophyseal hormones, but they may be secondarily induced. In the experiments by Benoit and Assenmacher (1953a,b, 1955) and Assenmacher (1957a,b), there is the possibility that they lesioned the infundibular nucleus and/or sectioned its axons proceeding to the median eminence. It has recently been demonstrated that the infundibular nucleus regulates gonadotropin release from the adenohypophysis.

John and George (1967) observed that the neurosecretory cells of the nucleus supraopticus and nucleus paraventricularis contain little neurosecretory material in the postmigratory period, but they are heavily loaded in the premigratory period. A similar tendency was observed in the amount of neurosecretory material in the median eminence and the neural lobe.

Functions of Neurohypophyseal Hormones

As mentioned before, both the pars nervosa and the median eminence contain the same neurohypophyseal hormones—arginine vasotocin and oxytocin. However, the ratio of vasopressor activity to oxytocic activity is higher in the median eminence. It is possible that arginine vasotocin may be released more actively than oxytocin from the median eminence into the portal vessels or that the median eminence receives more axons from the neurosecretory cells producing mainly arginine vasotocin than from those producing oxytocin. Following dehydration in the Japanese Quail, the hormone content of the hypothalamohypophyseal system showed a decrease. Arginine vasotocin showed a greater decrease (82.1%) than did oxytocin (57.9%). Arginine vasotocin in the median eminence decreased by 36.0%. These results showed that vasotocin is an active antidiuretic principle. Oxytocin is diuretic in birds.

It has been demonstrated that the neurohypophyseal hormones in the median eminence change in amount as the testes grow in the duck and in *Zosterops*. However, Follett and Farner (1966) found a change in one experiment, but could not find it in another experiment in the Japanese Quail. In those experiments, the hormones in the pars nervosa did not show any change in amount, and there were no conspicuous changes in the neurosecretory material. Here again, the relationship between neurosecretory system and gonadal development is not clear in terms of neurohypophyseal hormones. From these experiments, it is obvious that the neurohypophyseal hormones are not the main substance involved in gonadal growth, although vasopressin was once regarded as a substance that releases adenohypophyseal hormones. The change in the amount of the hormones in the above experiments may be induced secondarily.

One important finding is that the neurohypophyseal hormones in the pigeon median eminence, but not in the pars nervosa, decrease markedly after the systemic injection of formalin. It is suggested that the hormones in the median eminence might be related to ACTH release rather than to gonadotropin release. Supporting this idea, Kawashima et al. (1964) observed a decrease of sudanophilic materials in the adrenal gland with concomitant decrease of the neurosecretory material after water deprivation in the White-crowned Sparrow. Rossbach (1966) observed a correlation between adrenal cortical activity and the nucleus magnocellularis interstitialis dorsalis, which is neurosecretory. However, considering the presence of corticotropin-releasing factor in the hypothalamus, it is likely that the changes in the amount of the neurohypophyseal hormones and neurosecretory material may be secondarily induced by environmental changes.

Tanaka and Nakajo (1962) showed a marked decrease of arginine vasotocin, but a slight decrease of oxytocin, in the fowl neural lobe after oviposition. Sturkie and Lin (1966) found that the plasma vasotocin level rises remarkably at the time of oviposition. Single intravenous injection of arginine vasotocin induces oviposition within 90 seconds. Arginine vasotocin is more potent in causing contraction of the fowl oviduct than oxytocin. However, Shirley and Nalbandov (1956) found that neurohypophysectomy does not affect oviposition.

Although the biological significance is obscure, oxytocin induces an increase of glucose and fatty acids in the blood of the domestic fowl and of blood sugar in the White-crowned Sparrow. Oxytocin has a vasodepressor activity in the domestic fowl and other birds; and this phenomenon is used for the bioassay of the peptide.

Electron Microscopic Observations

Bern et al. (1966) studied the changes in the ultrastructure of the median eminence of the White-crowned Sparrow subjected to long days and found that there was no correlation between the type, size, or number of vesicles and gonadal growth. However, osmotic stress induced a decrease in the number of elementary granules.

As mentioned earlier, Peczely and Calas (1970) have observed four types of granules (1600-1900 Å, 1200-1400 Å, 1000 Å, and 600-800 Å) and synaptic vesicles in the median eminence of the pigeon. Adenohypophysectomy induced almost complete disappearance of electron-dense granules in the external layer of the anterior median eminence and induced an increase in empty axons and an increase of the granules (1600-1900 Å) in the posterior median eminence. Following the intravenous injection of insulin, the electron-dense granules in the anterior part of the anterior median eminence were largely depleted, but the granules (1200-1400 Å and 1600-1900 Å) were increased in the posterior median eminence. Metopirone caused the granules (1200-1400 Å) to increase in both anterior and posterior divisions of the median eminence and a significant increase in the number of synaptic vesicles in the posterior median eminence. Prednisolone induced a marked increase of the granules of 1200-1400 Å in the anterior median eminence and induced an increase of the granules of 1000 Å in the posterior median eminence. From all these findings Peczely and Calas (1970) concluded that the granules of 1200-1400 Å have an important role in releasing ACTH from the adenohypophysis. It was difficult to draw other conclusions from the changes in the number of granules of the different types. After hypophysectomy in the pigeon, large granules (1600-1900 Å) increased in number in the posterior median eminence. This phenomenon is in good agreement with our previous observations that AF-positive neurosecretory materials are accumulated in the posterior median eminence after hypophysectomy, where we rarely see AF-positive neurosecretory material. This finding suggests that the AF-positive neurosecretory material may be involved in some obscure function in the posterior median eminence. The changes in the population of different types of granules in the mammalian median eminence are reviewed by Kobayashi et al. (1970).

Enzymes in the Neurosecretory System

Kobayashi and Earner (1960) attempted to assess the activity of the neurosecretory system by measuring the activity of acid phosphatase, since this enzyme reflects the general metabolism of cells. It was found that the activity of acid phosphatase in the median eminence, but not in the pars nervosa, increased during testicular growth in the photosensitive White-crowned Sparrow subjected to long days. The supraoptic region did not show any significant change in enzyme activity following exposure to long days. The enzyme activity of the median eminence of refractory birds subjected to long days showed no increase. These phenomena indicate that the median eminence is involved in the regulation of gonadal growth, and that the median eminence and the pars nervosa function independently. Following dehydration, the acid phosphatase activity increased in the supraoptic region and the pars nervosa, but not in the median eminence region. Again, the independent functions of the eminential and neural lobe components were confirmed. The pigeon responded in a similar way to dehydration. The proteinase activity was also used for quantitative expression of neurosecretory system. The change in the activity of this enzyme showed a similar tendency to that of acid phosphatase following exposure of the White-crowned Sparrow to long days or to procedures leading to dehydration.

The White-throated Sparrow (*Zonotrichia albicollis*) showed more activity increase in the supraoptic region, median eminence, and adenohypophysis following exposure to long days than the White-crowned Sparrow. Thus, the level of acid phosphatase activity is a useful indicator of the activity of the parts of the hypothalamohypophyseal system. This technique was used for demonstration of a negative gonadal feedback on the hypothalamus. When castrated *Emberiza* and White-crowned Sparrows were subjected to long days, the phosphatase activity of the adenohypophysis increased more than in the intact birds. Administration of estrogen or androgen decreased the enzyme activity of the adenohypophysis. The increased enzyme activity of the adenohypophysis returned to the initial level earlier in the intact birds than in the castrated ones. These findings show that the gonadal feedback on the hypothalamus is operating, but weakly, during the period of gonadal development, and that the feedback is not a main factor for the termination of photoperiodically induced gonadal development.

Tanaka and Nakajo (1959) found that the activity of cholinesterase shows different levels at the times just before, during, and just after oviposition in the fowl. They suggested that in the laying hen the enzyme activity might be related to the stimulation of the oviduct by the egg. Kobayashi and Farner (1964) and Follett et al. (1966) found that the neurosecretory cells and the palisade layer of the median eminence show strong acetylcholinesterase (AChE) activity and that the pars nervosa shows weak enzyme activity in the White-crowned Sparrow. Uemura (1964c, 1965) obtained similar results in *Zosterops japonica*. These findings suggest that the neurosecretory neurons are cholinergic in nature, and that some cholinergic mechanism is functioning in the median eminence. The capillaries of the primary plexus showed no pseudocholinesterase activity, whereas those of the brain showed strong activity. The absence of enzyme activity in the capillaries of the primary plexus may be related to the active transport of the neurohormones through the membrane of the endothelial cells. Russell (1968) and Russell and Farner (1968) measured the enzyme activity in the hypothalamohypophyseal axis of the White-crowned Sparrow. They found that the acetylcholinesterase activity in the adenohypophysis increases during testicular development. There is a consistent daily cycle of AChE activity in the median eminence. Winget et al. (1967) measured the activity of acid phosphatase, alkaline phosphatase, and cholinesterase in the diencephalon, pituitary, and plasma of the domestic fowl exposed to light after darkness for 56 days. Illumination increased the activity of all enzymes except hypothalamic alkaline phosphatase and pituitary cholinesterase. They concluded that visible radiation is a *Zeitgeber* for the enzyme systems evaluated.

Matsui and Kobayashi (1965), in the Tree Sparrow (*Passer montanus*), showed the presence of a strong monoamine-oxidase reaction in the subependymal layer and the palisade layer, especially in the tissues near the capillaries of the primary plexus. This indicated that monoaminergic fibers terminate around capillaries. The neurosecretory cells gave a very weak monoamine-oxidase reaction. The distribution of monoamine oxidase (MAO) in the median eminence is different from that of neurosecretory material. Therefore, neurosecretory neurons are probably not monoaminergic. However, there is strong MAO activity around the perikarya, suggesting that the neurosecretory cells may be innervated by monoaminergic fibers. Results similar to the above were obtained by Follett et al. (1966) in the White-crowned Sparrow, and by Urano (1968) in the Japanese Quail. In both experiments, the infundibular nucleus showed a strong MAO reaction. The distribution of MAO reaction is very similar to that of monoamine fluorescence and also to that of small granules carrying monoamines. Follett (1969) measured quantitatively the activity of MAO of the median eminence and the basal hypothalamus of the Japanese Quail. He found that there is a daily rhythm in the enzyme activity, but he could not find a clear relationship between MAO activity and the release of gonadotropins resulting from photoperiodic stimulation.

Ependymal Function in the Median Eminence

In the domestic fowl, Legait (1959) observed that stainable material is secreted from the ependymal cells of the median eminence. However, this phenomenon was not observed in the White-crowned Sparrow and the Zebra Finch, although it was noticed that the ependymal cells of the median eminence have extended small protrusions into the third ventricle. Later, with an electron microscope, the detail of the ependymal protrusions was studied in the median eminence of the White-crowned Sparrow. The first type is the bleb-like microvillus or bulbous protrusion. It usually contains polysomes, glycogen granules, mitochondria, and small vesicles or vacuoles. Some of these protrusions are detached from the cell body as masses and others have ruptured membranes discharging their inclusions. The second type is the fingerlike microvillus, which rarely includes cytoplasmic organelles. At the cell surface near the base of these microvilli, small pinocytotic vesicles, seemingly impounding a droplet of cerebrospinal fluid, are often observable. This may be concerned with the absorbing function of the ependymal cells. The third type consists of the surface folds and the marginal folds. The marginal fold protrudes from the margin of the ependymal cell surface and the surface fold protrudes from any point of the apical surface. They recurve their free ends toward the cell surface and entrap a droplet of the ventricular fluid. Matsui (1966b), Oota (1970), and Galas and Assenmacher (1970) have also called attention to protrusions from the ependymal cell surface of the median eminence of the pigeon and duck. Recently, we have demonstrated that the ependymal cells of the median eminence of the Japanese Quail absorb peroxidase injected into the third ventricle. This is the first clear evidence showing the absorption of ventricular fluid by ependymal cells.

As mentioned above, the ependymal processes in the median eminence have synaptic contacts with monoaminergic fibers in the pigeon. It is probable that the ependymal secretion and absorption may be regulated by the monoaminergic fibers. The origin of the fibers is not known. The ependymal function in the mammalian median eminence is discussed in relation to the adenohypophyseal function in a review by Kobayashi et al. (1970). In the neural lobe of the domestic fowl, Payne (1959) observed that a part of the apical cytoplasm of the ependymal cells is secreted into the lumen.

Functional Aspects of the Gomori-Negative Hypophysiotropic Neurosecretory System

Hypothalamic Regions Regulating Adenohypophyseal Function

It is evident that in mammals afferent neural information is transformed into neurohormonal information (releasing factors) in certain Gomori-negative neurosecretory cells, and then the neurohormonal information is transported to the median eminence and conveyed to the adenohypophysis through the portal vessels, resulting in secretion of adenohypophyseal hormones. A similar series of events involved in hypothalamic control of the adenohypophysis seems to occur also in birds. Centers regulating adenohypophyseal functions in birds have not always been as clearly identified as in mammals. However, several attempts have been employed to assess exact sites regulating the release of adenohypophyseal hormones by means of lesions, electric stimulation, and hormone implantation in certain nuclei of the hypothalamus.

Gonadotropin

The hypothalamic control of the release of gonadotropin has been demonstrated in many experiments, such as those involving sectioning portal vessels and severing the median eminence. The first attempt to introduce lesions into the avian hypothalamus was carried out by Assenmacher (1957a,b, 1958) in the domestic duck. The lesions located in the supraoptic and paraventricular nuclei were always followed by genital atrophy with the same latency as after hypophysectomy (2-3 weeks). From the results and those obtained previously in Benoit's laboratory, they concluded that neural information leading to gonadotropin release came from the supraoptic and paraventricular nuclei to the median eminence and was conveyed to the adenohypophysis through the portal vessels. Supporting this idea, Egge and Chiasson (1963) found that electrolytic lesions in the nucleus paraventricularis, median eminence, and the pars tuberalis induced cessation of the egg-laying cycle and reduction in size of the comb and wattles in the domestic fowl. Since the lesions were made in the nucleus paraventricularis, it

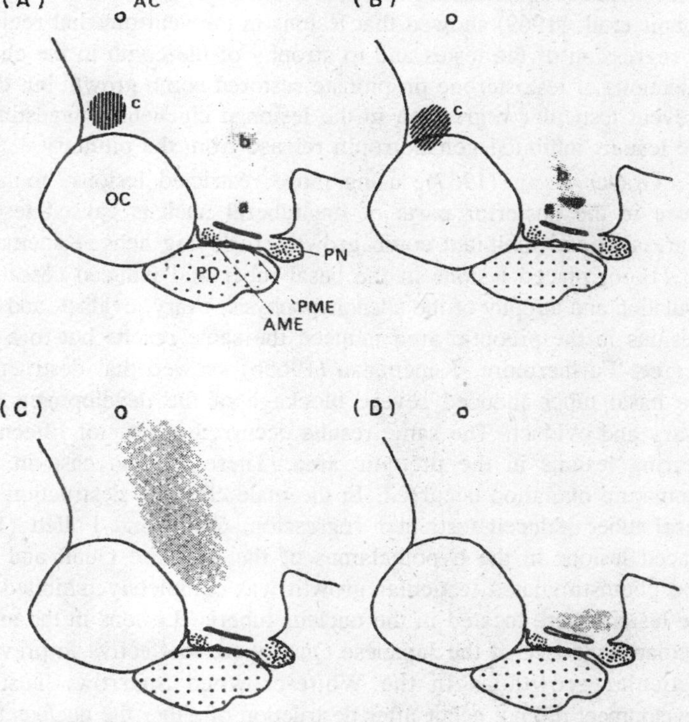

Fig. 2.4. Possible sites regulating adenohypophyseal function. Gonadotropin regulatory sites revealed by lesions: (a) nucleus tuberis; (b) nucleus hypothalamicus posterior medialis; (c) preoptic area. (B) Androgen sensitive sites: (a) nucleus tuberis; (b) site sensitive to testosterone in ducks; (c) sensitive site in duck where testosterone induced partial testicular regression. (C) Possible regulatory site of TSH release. (D) Possible regulatory site of ACTH release. AC, anterior commissure; AME, anterior median eminence; OC, optic chiasma; PD, pars distalis; PME posterior median eminence; PN, pars nervosa.

seems that AF-positive materials are responsible for gonadotropin release. As mentioned before, there have been many histological results supporting this idea in passerine birds. However, Wilson (1967) placed lesions in the hypothalamus of the White-crowned Sparrow and showed that lesions in the anterior median eminence, which is the depot of AF-positive material, gave no effect on photostimulated testicular growth, but lesions in the tuberoinfundibular region completely inhibited photoinduced testicular growth. Stetson (1969a) confirmed by lesioning experiments that the nucleus infundibularis is responsible for photostimulated testicular growth of the White-crowned Sparrow. Lepkovsky and Yasuda (1966) found that the destruction in the ventromedial region induced testicular atrophy in the chicken. Further, Snapir et al. (1969) showed that lesions in the ventromedial region led to regression of the testes and to atrophy of the comb in the chicken. Injections of testosterone propionate restored comb growth but did not prevent testicular regression in the lesioned chickens, suggesting that the lesions inhibited gonadotropin release from the pituitary.

Graber et al. (1967), using more restricted lesions, found that those in the posterior parts of the tuberal nucleus caused testicular regression and inhibited comb growth. In laying hens, Kanematsu et al. (1966) placed lesions in the basal tuber and induced cessation of ovulation and atrophy of the adenohypophysis, ovary, oviduct, and comb. Lesions in the preoptic area induced the same results but to a lesser degree. Furthermore, Kanematsu (1968b) showed that destruction in the basal tuber induced severe blockage of the development of the ovary and oviduct. The same results occurred in four of fifteen birds bearing lesions in the preoptic area. There was no case in which premature ovulation occurred. In the male chicken, destruction in the basal tuber induced testicular regression. Sharp and Follett (1969c) placed lesions in the hypothalamus of the Japanese Quail and found that photostimulated testicular growth was completely inhibited when the lesions were located in the nucleus tuberis. Lesions in the anterior median eminence of the Japanese Quail were ineffective in preventing testicular growth as in the White-crowned Sparrow. Testicular development did not occur after destruction of either the nucleus tuberis or the nucleus hypothalamicus posterior medialis. They found that there are monoaminergic fiber connections between both nuclei. Perhaps the latter nucleus regulates the former. Stetson (1969c) also reported that electrolytic destruction in the hypothalamus of adult breeding male and female Japanese Quail caused gonadal disfunction. Lesions in the rather posterior infundibular nuclear region seems to inhibit an LH-like hormone secretion in both sexes, and lesions in the anterior infundibular nuclear region seems to inhibit an FSH-like hormone secretion in both. The results suggest that in both male and female Japanese Quail, two separate regions appear to be involved in gonadotropin secretion.

Ralph (1959) found that electrolytic lesions in the ventral preoptic hypothalamus were consistently followed by an immediate cessation and prolonged interruption of ovulation in adult laying hens. Further, Ralph and Fraps (1959) showed that electrolytic destruction in the preoptic area inhibited progesterone-induced ovulation. Electrical stimulation of the anterior median eminence 14 hours before the next ovulation induced delayed ovulation. Opel (1963) stimulated the preoptic area and found that ovulation was postponed if stimulated 14 hours before the next ovulation. Ralph and Fraps (1960) showed that progesterone injected into the anterior hypothalamus and the preoptic area induced premature ovulation. Thus, the preoptic area seems to be involved in the mechanism of ovulation, and progesterone stimulates the preoptic area to induce ovulation. As for the oviposition, if stimulation was applied 4-25 hours before oviposition, premature oviposition was induced. He is of the opinion that premature oviposition occurred because of the release of vasotocin in this case. However, the neurohypophyseal hormones do not appear to be essential for oviposition by neurohypophysectomy.

Androgen implantation was carried out by several investigators to determine the sensitive sites based on the concept of a negative feedback. Kordon and Gogan (1964) found that implants of testosterone in the ventromedial nucleus of the domestic duck inhibit testicular growth by photostimulation. Gogan (1968) confirmed this, and further, that in addition to this nucleus, implants in the preoptic area partially inhibit testicular growth. In the latter case, spermatogenesis ceased at the stage of spermatocyte I, but the Leydig cells seemed to be active. Photostimulated testicular growth was inhibited in the [American] Tree Sparrow (*Spizella arborea*) by implantation of testosterone in the basal region of the nucleus infundibularis. In the Japanese Quail, the situation is the same as in the [American] Tree Sparrow. Implants of testosterone in the nucleus tuberis inhibited photostimulated testicular growth. Implants in other hypothalamic regions including the nucleus hypothalamicus posterior medialis of Sharp and Follett (1969c) did not give clear inhibition of photoinduced testicular development. Recently puromycin was implanted into various regions of the hypothalamus of the Japanese Quail. Only when implanted into the nucleus tuberis was puromycin effective in suppression of photoinduced testicular growth. It is known that the avian infundibular nucleus is topographically divisible into several cell groups. It is not known which cell groups were affected by puromycin.

From the results mentioned above, it could be concluded that the ventromedial region of the hypothalamus, or more precisely the nucleus tuberis, is the regulatory site of gonadotropin secretion from the adenohypophysis. In the hen, the preoptic area is responsible for gonadotropin release for ovulation.

Prolactin

Since section of the portal vessels induced atrophy of the prolactin cells in the duck, it is obvious that secretion of prolactin in birds is regulated by the hypothalamic prolactin-releasing factor. No direct evidence for the presence of a regulatory center of prolactin secretion is available.

Thyrotropin

Secretion of thyrotropin seems to be regulated by the hypothalamus, since the section of the portal vessels induced a decrease in the thyroid weight in the duck and the pigeon. Assenmacher (1958) reported that a certain number of ducks showed atrophy of the thyroid gland after lesioning of the anterior hypothalamus or complete sectioning of the median eminence at the level of just before the anterior median eminence. Egge and Chiasson (1963) found that lesions in the preoptic division of the nucleus supraopticus induced a decrease in thyroid weight. However, there was no apparent morphological difference in the follicular epithelium between the birds with light and heavy thyroid glands, except that the birds with lesions lacked the reabsorption lacunae in the central region of the thyroid glands. McFarland et al. (1966) lesioned the anterior hypothalamus including the ventrolateral nuclei and induced a decrease in the thyroxine disappearance rate in the Japanese Quail. Lesions in the basal anterior hypothalamus, which includes the nucleus hypothalamicus posterior medialis, induced atrophy of the thyroid gland in cockerels. Another type of experiment has been carried out to assess the hypothalamic regulating site of thyrotropin secretion in the chicken. Adenohypophyseal tissue transplanted in the area near the nucleus paraventricularis magnocellularis showed hypertrophy of the basophils after thyroidectomy. Appearance of thyroidectomy cells in the grafted adenohypophysis was inhibited by destruction of the anterior hypothalamus located ventrocaudal to the anterior commissure and dorsocaudal to the optic chiasma. Only four of thirty-nine chickens showed a markedly low thyroidal uptake of ^{131}I. The lesion in these birds was found in the median eminence, except in one bird in which the lesion was located in the nucleus paraventricularis magnocellularis. The protein-bound ^{131}I conversion ratio in the cockerel gave clearer results. Protein-bound ^{131}I was depressed when the region ventrocaudal to the anterior commissure was destroyed. The region includes the nucleus paraventricularis magnocellularis. Destruction of the anterior hypothalamus, which includes the nucleus hypothalamicus posterior medialis, and of the region dorsocaudal to the optic chiasma also depressed the protein-bound ^{131}I conversion ratio. The thyroid gland decreased significantly in weight after destruction of the anterior hypothalamus, including the nucleus hypothalamicus posterior medialis and the nucleus paraventricularis.

Although the evidence obtained so far is not sufficient to draw a conclusion about the exact regulatory site of thyrotropin secretion, the anterior hypothalamus, including the nucleus paraventricularis magnocellularis and the nucleus hypothalamicus posterior medialis, seems to be the site responsible for the control of thyrotropin secretion from the adenohypophysis. Thus, the regulating site is not confined to one nucleus.

Corticotropin

Section of the hypophyseal portal vessels causes a decrease in adrenal weight and in the ratio of cortical to medullary tissue in sectioned birds. However, the effect of severance of portal vessels is not the same as that of adenohypophysectomy. Boissin (1967) sectioned the portal vessels of the duck and showed that adrenal weight and the ratio of cortical to medullary tissue decreased significantly; the decrease was greater in the birds with unregenerated portal vessels than in those in which the vessels were restored. In the sectioned birds, adrenal ascorbic acid did not significantly decrease, and plasma corticosterone concentration of the peripheral blood was not significantly affected. However, corticosterone concentration in the birds with the unregenerated portal vessels had a tendency to be lower than that in the birds with regenerated portal vessels. In the chicken, autotransplantation of the adenohypophysis under the kidney capsule reduced the plasma corticosterone concentration in the adrenal vein to the same level as that of adenohypophysectomized subject. The peripheral concentration of plasma corticosterone did not differ significantly from that of intact birds after autotransplantation in the duck, pigeon, and Japanese Quail. The differences in results among investigations may be due to the difference in species or in experimental period. Resko et al. (1964) killed their chickens 40 days after the implantation and Bayle et al. (1971) killed their ducks 63 days after the implantation.

Miller (1961) found that the adrenals atrophied after hypophysectomy in the pigeon but showed markedly hypertrophic cortical tissue after treatment with formaldehyde or insulin. After large lesions in the pigeon median eminence region including the ventral hypothalamus in addition to total hypophysectomy, the adrenals became hypertrophied rather than atrophied, and the response of these pigeons to formaldehyde was the same as that in hypophysectomized pigeons without lesions. Miller proposed three possible explanations: (1) There may be another source of corticotropin that was not destroyed in this experiment; (2) there may be a stimulus other than corticotropin that evokes adrenal enlargement; (3) the effect of corticotropin may not be direct on the adrenal but mediated via some pathway that can also be aroused by stress and lesion in the median eminence. Egge and Chiasson (1963) placed lesions in the hypothalamus of hens and concluded that no correlation exists between the lesioned area and the corticoid titer in the adrenals. Frankel et al. (1967a) concluded from the results obtained with hypothalamic lesions and adenohypophysectomy that the release of corticotropin from the adenohypophysis is controlled by the hypothalamus, and especially by the ventral part of the tuberal nucleus. Further, they found that the adrenals are stimulated not only by the adenohypophyseal corticotropin but also by an extrahypophyseal corticotropin-like substance, since lesions in the ventral part of the tuberal nucleus of the hypophysectomized cockerel evoked a higher corticosterone concentration than that in intact lesioned cockerels.

Plasma corticosterone level in the adrenal venous blood of the adenohypophysectomized cockerel decreased immediately after the surgery to about one-third of the level found in intact birds and this level was maintained up to 42 days later. However, the corticosterone level in ducks declined steadily during the first 2 weeks after adenohypophysectomy and then remained at a level that was one-third that of intact birds. Experiments show that adenohypophysectomized pigeons can respond to stress. Since the plasma corticosterone concentration remained one-third of that of intact birds after adenohypophysectomy, it is possible that the adrenal gland functions autonomously or that there is extrahypophyseal corticotropin. The latter possibility is more probable than the former, since ACTH or ACTH-like activity has been found in the median eminence or hypothalamus. More recently, Bradley and Holmes (1971) have reported that the peripheral concentration of corticosterone falls to only 10% of that of sham-operated ducks 14 days after adenohypophysectomy. Thus, adrenal cortical tissue is greatly dependent on the adenohypophysis. The authors are of the opinion that it is not necessary to postulate either a high degree of adrenal autonomy or an extrahypophyseal corticotropin source to maintain normal adrenocortical function in the duck.

As mentioned above, there are differences of opinion among investigators with respect to the autonomous function of cortical tissue, its dependency on adenohypophysis, and extrahypophyseal corticotropin. Nevertheless, it is obvious that cortical function is largely regulated by the hypothalamohypophyseal system.

Growth hormone

There seems to have been no experimental attempts to establish the hypothalamic site that controls the release of growth hormone.

Releasing Factors and Monoamines

Extensive investigations have been carried out in mammals, and now the presence of the releasing or inhibiting factor for each adenohypophyseal hormones has been demonstrated. The following hypophysiotropic neurohormones are reported: luteinizing hormone-releasing factor (LRF), follicle stimulating hormone-releasing factor (FRF), prolactin-inhibiting and -releasing factors (PIF, PRF), thyrotropin-releasing factor (TRF), growth hormone-releasing and -inhibiting factors (GRF, GIF), and melanocyte stimulating hormone-releasing and -inhibiting factors (MRF, MIF). Recently, it was found that the active moiety of TRF is composed of only three amino acid residues and that LRF, which also has intrinsic FRF activity, contains ten residues. These results indicate that the releasing factors are probably small peptides. In birds, although few efforts have been made on this subject, the presence of some releasing factors have been demonstrated.

Gonadotropin-releasing factors

In birds, it is as yet open to question whether two types of gonadotropin (LH and FSH) are separable or not, although cytologically there appear to be two kinds of gonadotropic cells in the avian adenohypophysis. Recently, Hartree and Cunningham (1969) effected a partial purification of the chicken pituitary hormones by chromatography and found two fractions, one having FSH activity and the other LH activity. Accordingly, there may be separate releasing factors (LRF and FRF) for LH and FSH.

The first evidence suggesting the presence of LRF is the fact that infusion of hypothalamic extracts into the pituitary of laying hens induces premature ovulation. Cortical extracts and neurohypophyseal extracts were ineffective. Synthetic oxytocin and arginine vasotocin were also ineffective. On gel filtration (Sephadex G-25) of stalk median eminence, the peak of ovulating hormone-releasing activity was found to overlap with the peak of oxytocin and vasotocin as determined by hen and rat uterine assays. Infusion of 500 μg of lyophylized powder from this fraction induced ovulation in 13 of 21 hens. *In vitro*, hypothalamic extracts induced the release of LH from cock pituitaries. Acetic acid extracts from the hypothalamus of White Leghorn hens (1-2 years of age) were centrifuged and the supernatant was immersed in boiling water for 10 minutes to inactivate the LH that might be present as a contaminant in the extracts. This hypothalamic extract could induce LH release from the pituitaries of White Leghorn cocks incubated in Krebs-Ringer phosphate buffer (pH 7.4) containing 10 mM glucose. Cortical extracts did not induce LH release. Hypothalamic extracts of cockerels induced release of LH from rat pituitaries incubated *in vitro*. In this experiment, hydrochloric acid extracts of the hypothalamus of 12-week-old cockerels were centrifuged and the supernatant was used. The extracts induced significant LH release from the rat pituitary incubated in Medium 199. Arginine vasotocin in a concentration equal to that in the extracts had no effect on LH release under the same conditions. In all these experiments, the ovarian ascorbic-acid depleting (OAAD) method was employed for determination of LH. It has been demonstrated that chicken hypothalamic extracts contained another OAAD factor. This effect was not due to the presence of LH, but to arginine vasotocin in the extracts. Ishii et al. (1970) showed that OAAD activity in the pigeon median eminence is explained by arginine vasotocin therein. The activity was observed in the anterior median eminence where arginine vasotocin was localized. Recently, Jackson (1971a,b) partially purified chicken luteinizing hormone-releasing factor (LRF). Furthermore, he separated chicken LRF from arginine vasotocin and found it is chemically different from mammalian LRF.

Chicken hypothalamic extracts also stimulated FSH release from cock and rat pituitaries incubated *in vitro*. Hypothalamic extracts of turkeys also induced FSH release from the incubated pituitary tissue compared with cerebral cortex extracts. FSH was assayed by the Steelman-Pohley method.

Follett (1970a) demonstrated the presence of a gonadotropin-releasing factor in the hypothalamus of the Japanese Quail by an assay method other than OAAD. Hypothalamic extracts of the Japanese Quail were centrifuged, and the supernatant was added to Medium 199 that contained cockerel pituitaries. After incubation, the gonadotropic activity contained in the incubation media was determined by measuring the uptake of ^{32}P by the testes of 1-day-old chicks after injection of the media. Hypothalamic extracts enhanced gonadotropin release while cortical extracts did not. Extracts of the pars nervosa, synthetic vasopressin, and oxytocin did not evoke gonadotropin release from the incubated cockerel pituitaries. It is interesting to note that gonadotropin-releasing activity in the quail hypothalamus varies in accordance with reproductive states. The birds held under a nonphotostimulatory short daily photoperiod did not show detectable gonadotropin-releasing activity, whereas the birds under a long daily photoperiod showed considerable activity. There was no difference between the intact long-day males and castrated long-day males. Testosterone-implanted long-day males possessed high gonadotropin-releasing activity in the hypothalamus, although the testes of these birds were small and the gonadotropin content of the adenohypophysis was low. It is concluded from these experiments that in avian species gonadotropin-releasing factor(s) are present in the hypothalamus.

Prolactin-releasing factor

The control of prolactin release in birds differs from that of mammals. When the pigeon adenohypophysis was incubated for 6 days in vitro, there was no significant increase in the prolactin content of the incubation media. Nicoll (1965) examined the effects of acid extracts of cerebral and hypothalamic tissues of the Tricoloured Blackbird (*Agelaius tricolor*) on prolactin secretion from *Agelaius* adenohypophysis *in vitro*. The hypothalamic extracts induced prolactin secretion, indicating that the hypothalamus of *Agelaius* contained a stimulating factor for prolactin secretion. Cerebral cortex extracts had no effect on prolactin secretion. Kragt and Meites (1965) have demonstrated that hypothalamic extracts of pigeons stimulate the pigeon pituitary to secrete prolactin into the incubation media. Meites and Nicoll (1966) also showed that hypothalamic extracts from the chicken and quail increased prolactin release from the pigeon pituitary incubated *in vitro*. Gourdji and Tixier-Vidal (1966) demonstrated prolactin-releasing activity in the hypothalamus of the duck. Prolactin-releasing activity has also been demonstrated in the turkey hypothalamus *in vitro*.

Prolactin-releasing activity in the hypothalamus was much higher in the parent pigeons on the clay of hatching than in 4- to 6-week-old pigeons, suggesting that the prolactin-releasing factor is present in different concentrations in the hypothalamus at the different reproductive states. This phenomenon coincides with the observations on prolactin cells in pigeons that are feeding their young. In the case of the turkey, although total prolactin content in the adenohypophysis of 20-day-old male and female poults was much less than in hens killed at the end of the laying season, their hypothalami appeared to contain as much prolactin releasing activity as the hens.

Thus, in avian species, prolactin secretion appears to be controlled primarily by a prolactin-releasing factor in the hypothalamus, unlike a prolactin-inhibiting factor in mammals.

Corticotropin-releasing factor

As mentioned above, release of corticotropin in birds has been suggested to be controlled by the hypothalamus. Recently, it has been demonstrated that a corticotropin-releasing factor is present in the avian hypothalamus, as is adrenocorticotropic hormone (ACTH) or ACTH-like activity in the hypothalamus.

Peczely and Zboray (1967) demonstrated *in vitro* a corticotropin-releasing factor in the median eminence of the pigeon. They incubated the pigeon adenohypophysis with pigeon median eminence tissue homogenate. The incubation medium was then added to the other incubation media containing the rat adrenal tissue. Corticosterone secretion from the rat adrenal was enhanced. This suggests that the median eminence homogenate contained CRF and induced release of ACTH from the pigeon adenohypophysis into the incubation medium. In addition to CRF, they found that the pigeon median eminence contains more ACTH or ACTH-like substance than the adenohypophysis. Peczely et al. (1970) have demonstrated in the pigeon that ACTH-like activity in the median eminence is still as high as controls 1 month after adenohypophysectomy.

Stainer and Holmes (1969) demonstrated corticotropin-releasing activity in the hypothalamus of the duck, according to the following method. First, brain tissue of the duck was extracted and lyophylized, suspended in Krebs-Ringer bicarbonate buffer, and then added to the incubation media containing duck pituitary. Then the incubation medium was added to the other incubation media with slices of the duck adrenal gland. After incubation, corticosterone liberated in the media was determined by fluorometry. This CRF activity was also present in extracts of the cerebrum and spinal cord. ACTH or ACTH-like activity was not detected in the duck hypothalamus, unlike the pigeon. More recently, Salem et al. (1970a,b) injected hypothalamic extracts of chickens into dexamethasone-treated rats. To ascertain whether the extracts induced ACTH release from the rat pituitary, depletion of adrenal ascorbic acid (AAAD) was examined in rat adrenal glands. Using this method, CRF was found in the hypothalamus of the chicken, but not in the cortex. In addition to corticotropin-releasing activity, chicken hypothalamus contained ACTH-like activity. The ACTH-like activity was demonstrated by measuring AAAD of the adrenals of hypophysectomized rats injected with the hypothalamic extracts. They calculated that ACTH-like activity contributed to the total AAA depleting activity of the chicken hypothalamus by 27-34%. The ACTH-like activity was not due to ACTH contamination, because the substance showing ACTH-like activity was heat resistant and thought to be a substance similar to α-MSH. CRF activity of the hypothalamus was not due to vasotocin, although vasotocin has ACTH-releasing activity, when used in large amounts. They calculated the relative AAAD activity in the chicken hypothalamus as follows: CRF, 73.5%; vasotocin, 5.0%; ACTH-like substance, 21.5%. Boiling for 1 hour completely destroyed the CRF activity. This is different from the CRF found in mammals. Using an intrapituitary microinjection technique, Sato and George (1972) detected CRF activity in the pigeon hypothalamic median eminence and showed a diurnal rhythm in the CRF activity.

Growth hormone-releasing factor

Muller et al. (1967) have demonstrated growth hormone-releasing activity in the pigeon hypothalamus by measuring the decrease of growth hormone content in the rat pituitary after systemic injection of pigeon hypothalamic extracts.

Monoamines

As mentioned earlier, many papers have appeared dealing with monoamine distribution in the hypothalamus. The high concentration of monoamines in the nucleus infundibularis and median eminence suggests that monoamines could perform some role in regulating adenohypophyseal functions. Systemic injection of reserpine to reduce the monoamine level in the hypothalamus inhibits testicular development in the duck, pigeon, and chicken, but not in the White-crowned Sparrow and Japanese Quail. Implantation of reserpine into the nucleus tuberis exerts no effect on photoperiodic testicular response in the Japanese Quail. Neither noradrenaline nor dopamine induces gonadotropin release *in vitro*. The physiological meaning of monoamines in the hypothalamus awaits further experimental elucidation.

Concluding Remarks

Considering the data collected so far, it may be stated that a great deal of interest has been uncovered, but there is a great deal more to be done. Significant problems in avian neuroendocrinology still to be resolved include the following.

1. There appear to be functional differences among the clusters of neurosecretory cells in the neurosecretory nuclei. The functions of these individual clusters of cells have not yet been determined.
2. There are monoaminergic fibers and strong monoamine oxidase activity around the perikarya of the neurosecretory cells. Electron microscopy has revealed that they are innervated by monoaminergic fibers. Is the monoaminergic fiber the only one that communicates with the neurosecretory cell?
3. Do the neurosecretory cells function as osmoreceptors?
4. The median eminence and the neural lobe function independently, although they receive neurosecretory fibers of apparently common origins. What is the mechanism involved in this mutual independence?
5. What are the physiological significances of neurohypophyseal hormones and monoamines in the median eminence?
6. The physiological significance of the ependymal secretion into the third ventricle and the ependymal absorption of the ventricular fluid is not known at the present time.
7. What is the role of the thick perivascular space of the primary plexus? Does it function as a reservoir for releasing factors?
8. In birds, nothing is known about the chemistry of releasing factors. Where are the sites of production of the factors in the hypothalamus? How are the producing sites related to the regulating centers found by lesioning and by steroid implantation?
9. How are the production and the secretion of releasing factors regulated?
10. Are there regionally differentiated areas in the median eminence, each of which, containing a different releasing factor, corresponds to the differentiated areas of different cell types in the adenohypophysis?

11. Are all the releasing factors associated with granules or vesicles in the median eminence? If so, which types of granule contain the various releasing factors?

In the avian hypothalamus, there are two neurosecretory systems —fuchsinophilic (aldehyde fuchsin-positive) and nonfuchsinophilic. The former consists of the following three parts: (1) the cells of the nucleus supraopticus and the nucleus paraventricularis, in which neurohormones such as arginine vasotocin and oxytocin are produced; (2) their axons through which the neurohormones are transported; and (3) their axon terminals in two neurohemal regions—the anterior median eminence and the pars nervosa. Arginine vasotocin and oxytocin occur in both neurohemal regions. They are released from the pars nervosa into the systemic blood, and from the median eminence into the portal vessels. The release of these hormones seems to be controlled by monoaminergic fibers, since many fibers revealed by monoamine oxidase histochemistry and fluorescence microscopy are in contact with perikarya of the AF-positive neurosecretory cells. Arginine vasotocin is antidiuretic and vasopressor, and oxytocin is diuretic and vasodepressor. Arginine vasotocin is more potent in contractile actions on the fowl oviduct than oxytocin. Oxytocin induces increases of fatty acids and blood sugar in at least some species of birds. The biological significance of the neurohypophyseal hormones in the median eminence is not clearly known in relation to functions of the adenohypophysis. Although evidence is lacking, it is possible that this system produces some adenohypophysiotropic neurohormones that reach the pars distalis via the anterior median eminence and the anterior portal vessels.

The AF-negative (nonfuchsinophilic) neurosecretory system consists of cells whose localizations are not entirely known, but has axons and the axon terminals in the median eminence. The neurohormones produced in this system are adenohypophysiotropic (releasing factors) and reach the pars distalis and pars tuberalis via the adenohypophyseal portal system.

The avian median eminence is clearly divided into two regions—the anterior and posterior divisions. Both divisions are composed of five layers: (1) ependymal, (2) hypendymal, (3) fiber, (4) reticular, and (5) palisade. Some AF-positive fibers arising both from the nucleus supraopticus and the nucleus paraventricularis pass into the fiber layer toward the pars nervosa, forming the supraopticohypophyseal tract. Other AF-positive fibers descend to the palisade layer of the anterior median eminence. Aldehyde fuchsin-negative fibers, as revealed by electron microscopy, terminate in both the anterior and posterior median eminences. Their terminals are located near the capillaries of the primary plexus of the hypophyseal portal vessels. Electron microscopy shows several types of granules and vesicles with different diameters in the AF-positive and AF-negative fibers of the median eminence. Large granules of 1500-2000 Å may be carriers of the neurohypophyseal hormones, and the granules of 1000 Å seem to be carriers of monoamines. The large granules are abundant in the anterior median eminence but are rare in the posterior median eminence. The granules of 1000 Å are present both in the anterior and posterior median eminences. Other granules are possible carriers of releasing factors. The perivascular space of the capillaries of the primary plexus sends more digitations into the parenchyma of the posterior median eminence than into that of the anterior median eminence.

The ependymal cells of the median eminence seem to secrete some substance into the third ventricle and absorb some material from the third ventricle. These ependymal cells extend their processes to the capillaries of the primary plexus. The presence of the ventriculo-hypophyseal system is suggested in relation to hypothalamic control of the adenohypophysis.

In the adenohypophyseal hormone-releasing factor system, the tuberoinfundibular component seems to have an important role in gonadotropin secretion. Some of the cells of the nucleus infundibularis are monoaminergic and others are not. Both neurons send their axons to the palisade layer of the median eminence. These fibers form the tuberoinfundibular tract. The roles of the monoamine neurons are not known. The nonmonoaminergic neurons may produce gonadotropin-releasing factors. Destruction of the basal portion of the nucleus infundibularis interrupts gonadotropin secretion. Testosterone implanted in this nucleus interferes with gonadotropin secretion, resulting in atrophic testes. The preoptic area seems to be involved in ovulation. There is now some evidence for the occurrence of gonadotropin-releasing factors, a prolactin-releasing factor, an adrenocorticotropin-releasing factor, and a growth-hormone-releasing factor in the avian hypothalamus. These are released into the portal vessels and conveyed thereby to the adenohypophysis. It is not known, however, which neural components are responsible for the production of respective releasing factors; nor is the chemical nature of the releasing factors known.

3

PERIPHERAL ENDOCRINE GLANDS

The endocrine glands are essential components of the complex neuroendocrine apparatus that, by communication via nerve impulses, synaptic transmitters, and hormones, provides the basis for internal regulation and adjustment to the changing environment. The artificial designation "peripheral endocrine glands," is used here for those endocrines that are neither a part of the brain, such as the hypothalamus, its derivative the neurohypophysis, and the pineal organ, nor closely associated with it, i.e., the pars distalis and the pars tuberalis of the hypophysis.

The avian peripheral endocrine glands are, in general, typical of those of the higher vertebrate scheme, although they are by no means without specialized functions and adaptations that are typically avian. The high calcitonin content of avian ultimobranchial glands and its high specific activity seem related to the very intense calcium metabolism that appears characteristic of avian species, possibly linked to their high calcium requirement during the reproductive period. On the other hand, the very low binding capacity of avian plasma proteins to the circulating thyroid hormones, and the resulting short biological half-life of these hormones in birds, has been related to the high rate of heat production and to the high body temperature, another characteristic of avian physiology, that preadapts them better than mammals, for instance, to live in areas of intense heat. Other adaptations may be observed in special groups. The adrenal size, probably associated with a higher hormonal activity, of marine bird species, as compared with freshwater or terrestrial species, has been claimed to reflect the need of marine birds to stimulate chronically the extrarenal (nasal) salt excretory pathway. Other avian endocrinological peculiarities (e.g., the overwhelming lipolytic role of the pancreatic hormone glucagon versus the adrenomedullary catecholamines) are of unknown evolutionary or adaptative significance.

Adrenal Glands

Morphology

The adrenals are a pair of oval, pear-shaped, or triangular glands, yellow or orange in colour, that lie just anterior to the postcaval vein and to the cephalic lobe of the kidney. The avian adrenal gland, like that of most higher vertebrates, is formed of two components, the cortex (interrenal tissue) and the medulla (chromaffin tissue), which are of different origins and distinct functions.

The cortical portion is the first to be differentiated. The primordium buds off from the coelomic epithelium, ventrally and medially to the mesonephros. Soon after their formation, the prospective cortical cells leave the epithelium and move dorsally to form paired masses of scattered cell groups in the mesenchyma on each side of the aorta. During their further development, the cortical cells become arranged in cords, and blood cells and vessels appear between the cell cords. Medullary (chromaffin) cells appear several days later, arising from the primordium of the sympathetic nervous system. These cells migrate singly, ventrally between the aorta and groups of cortical masses, collect, and tend to arrange themselves in cords between the cortical tissue, before separating into groups of variable size around blood vessels.

In the adult animal, the ratio of cortical to chromaffin tissue is about 2 to 1. In most species, the cortical tissue is arranged in strands, which are surrounded by a highly vascular connective tissue. In longitudinal section, each strand appears as a double layer of columnar epithelial cells with the thin long axes perpendicular to the surface of the cell strand. Unlike the mammalian cortical cells, the avian interrenal cells have a marked polarity in the distribution of cytoplasmic organelles. The majority of lipid droplets, which are typical of cortical cells, and of the smooth-surfaced reticulum, occur in the basal cytoplasm adjacent to the connective tissue, whereas the Golgi area, dense bodies, and specialized attachment structures occur apically in the nuclear region.

Stimulation of cortical tissue, e.g., by ACTH injections, induces marked depletion of the lipid droplets, as classically observed with the light microscope. The lipid depletion also is evident at the electron-microscope level, together with (1) increased density of mitochondria; (2) increase of smooth-surface reticulum (SER), which is diffusely dispersed in the cytoplasm separating the mitochondria from the lipid droplets; (3) increase of the rough ergastoplasmic reticulum, which is absent in nonstimulated cells; (4) marked increase in size of the Golgi apparatus, associated with dense bodies. Correlatively, the nucleus and nucleolus increase in size, and numerous vacuoles appear at the surface of the nucleolus. Finally, numerous coated vesicles and invaginations appear, along with a fibrillar substance in intercellular spaces, as signs of uptake of material (proteins) into the cell. From a cytofunctional point of view, it is generally accepted that the close relationship among lipid droplets, SER, and mitochondria points to the basal cytoplasm as the main site of steroidogenesis, while other functions, such as protein synthesis, required by an increased need in enzymes involved in steroidogenesis, increased production of primary lysosomes, and increased degradation of lipids, may be attributed to the apical structures.

Another peculiarity of the arrangement of avian cortical tissue is a much greater homogeneity in cell population. Early investigations did not recognize any zonation in the avian adrenal cortex. More recently, several authors have described two different zones, a thin subcapsular zone and a thicker inner zone, e.g., in the pigeon, the Brown Pelican (*Pelecanus occidentalis*), and the duck. The two zones were easily demonstrated at the ultrastructural level in the fowl. From a histochemical standpoint, a thorough exploration of a number of species seems to indicate that the glands of certain birds are fully zonated (e.g., the pigeon), while others are completely homogeneous (e.g., Little Cormorant, *Phalacrocorax niger*). Similar variations have been observed in respect of the histological organization of the interrenal tissue in 22 species of 17 different orders. On the other hand, the subcapsullar layer seems particularly reactive to experimental saltwater balance alterations, and species that consume water in saline areas and on the seashore have been found to have less active peripheral cortical cells than do freshwater-adapted species.

In contrast with the cortical cells, the medullary chromaffin cells have no definite pattern of arrangement. They may be intimately intermingled with cortical tissue (e.g., in Passeriformes) or more or less condensed in medullary islets, intermixed with interrenal tissue (e.g., in Galliformes).

When studied with the electron microscope, two medullary cell types have been described in the domestic fowl and in the Gentoo Penguin (*Pygoscelis papua*), which differ in the size and shape of electron-dense granules. One cell type contains granules of very irregular size (800-5000 Å) and shape, while the other type is characterized by regular, spherical granules, that are also smaller than the others. The former are considered to correspond to norepinephrine cells, the latter cell type being epinephrine cells. On the other hand, the cells with irregular granules have been claimed to form the greater part of the medulla in the fowl, while an opposite ratio was observed in the Gentoo Penguin. Both types of granules consist of an electron-dense, central or eccentric osmiophilic deposit, which appears as a mass of numerous fine grains surrounded by a light halo, and finally by a limiting membrane. These granules, which seem to contain catecholamines, lie close to the Golgi apparatus, and it has been suggested that the outer membrane of the granules may originate from Golgi vesicles. Sometimes the limiting membrane of the granules contacts the cell membrane, and their content seems to be excreted through the opening of this contact zone, into the intracellular or perisinusoidal spaces. Between the medullary cell columns, there are bundles of unmyelinated nerve fibers invested by Schwann cells. At the contact zone between the nerve endings and medullary cells, a typical vesicular component, consisting of synaptic vesicles, is observed. Strong stimulation of medullary cells (e.g., by insulin) leads to a remarkable reduction in number, size, and electron density of the granules, with many smooth-surfaced vacuoles containing only diffuse fine micrograpules appearing in most cells.

Cortical Tissue

Hormones and metabolism

Corticosterone, as the main corticosteroid secreted by the avian adrenal, was first demonstrated in the efferent adrenal blood of castrated chickens, and of intact fowl, pheasants, and turkeys, and thereafter in a number of other species. Plasma corticosterone can be routinely measured by fluorometry, which is currently applied to mammalian species. However, in most avian species, except for ducks, a preliminary chromatographic purification of the plasma extract is necessary in order to obtain accurate assays. Aldosterone, another important avian corticoid, has also been shown to occur in adrenal efferent blood. The latter authors, and others also have identified small amounts of cortisol in the adrenal and peripheral blood of pullets, while Nagra et al. (1960) were unable to detect 17-oxycorticosteroids in the adrenal effluent blood from chicken, pheasant, or turkey. Corticosterone and aldosterone were also measured by the double isotopic dilution method in adrenal extracts from ducks, pigeons, and Japanese Quail. No significant interspecific variations could be detected, either in the concentration of both hormones or in the corticosterone to aldosterone ratio, which is approximately 8:1. Using the same accurate method, Daniel (1970) has shown that cortisol is present in the duck adrenal at a very low level (about 1.3 ng per gram of adrenal tissue).

A deeper insight into the metabolic sequences of the biogenesis of corticosteroids by the avian adrenal has been obtained from *in vitro* studies, which have been performed in a number of bird species. First, it has been confirmed that the avian adrenal synthesizes almost exclusively 17-deoxycorticosteroids, i.e., corticosterone, 18-hydroxycorticosterone, and aldosterone. Further investigations, using labeled presumed precursors, have shown that the metabolic pathways of steroidogenesis seem to conform to the classic route of other vertebrates: acetate → cholesterol → pregnenolone → progesterone → 11-deoxycorticosterone (DOC) → corticosterone → 18-hydroxycorticosterone (18 OH-B) and aldosterone. However, several peculiarities of the avian biogenesis of corticosteroids *in vitro* have recently been demonstrated with duck adrenal slices or intracellular fractions:

1. While exogenous [^{14}C] progesterone, as well as [^{3}H] pregnenolone, are easily converted to the three main corticosteroids, significant accumulation of labeled progesterone never occurs if [^{3}H]-pregnenolone is used. A metabolic sequence originating from pregnenolone, but without an oxidation step to progesterone, thus seems to exist, at least in the duck adrenal.
2. It has also been proved that, while both progesterone and 11-deoxycorticosterone are easily transformed to corticosterone, the hydroxylation at C-11 of progesterone might precede that of C-21, i.e., 11-deoxycorticosterone is not a necessary intermediate between progesterone and corticosterone. Duck adrenals convert either pregnenolone or progesterone to 11β-hydroxyprogesterone, which in turn leads to corticosterone without any production of 11-deoxycorticosterone. Corticosterone is further transformed to 18-OH-B and aldosterone, both hormones being also obtained from incubations with either progesterone or 11β-hydroxyprogesterone.

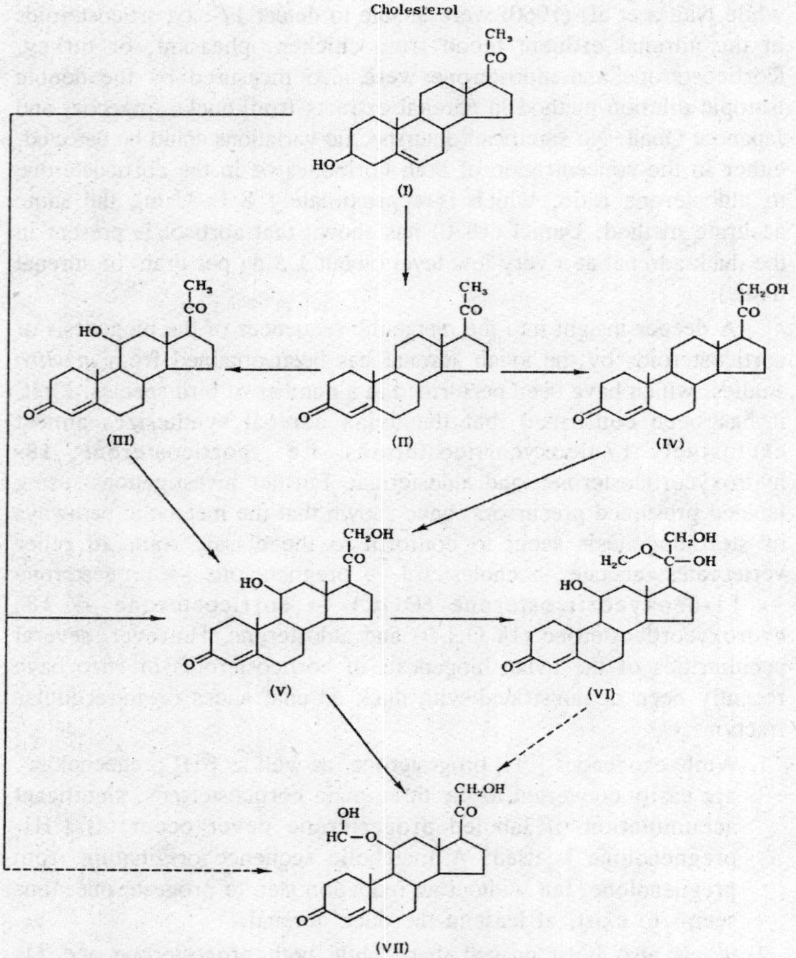

Cholesterol

Fig. 3.1. Proposed biosynthetic route of corticosteroid synthesis in the adrenal gland of the domestic duck. (I) pregnenolone; (II) progesterone; (III) 11-hydroxyprogesterone; (IV) 11-deoxycorticosterone; (V) corticosterone; (VI) 18-hydroxycorticosterone; (VII) aldosterone.

3. The synthesis of aldosterone from corticosterone is always accompanied by the simultaneous production of 18-hydroxycorticosterone. However, the question whether the latter is to be considered an obligatory precursor of aldosterone is still open since it has never been demonstrated to be converted by duck adrenal slices to aldosterone.

As in mammals, the disappearance rate of corticosterone from blood is rapid. In the duck, studies on the disappearance rate from plasma of [^{3}H] costicosterone indicate a biological half-life of about 11 minutes. The corresponding calculated metabolic clearance rate, which provides a better insight into the peripheral metabolism of the hormone, is 104 liters per day or 26.7 ml min^{-1} kg^{-1}. With the assumption that during the short time of the experiment the animals were in steady state with respect to corticosteroid metabolism, the secretion rate of the hormone is given by

secretion rate = metabolic clearance rate × plasma corticosterone

i.e., in the former example, 26.7 × 0.115 = 3.1 μg min^{-1} 'kg^{-1}. In similar experiments on pigeons Chan et al. (1971) found a metabolic clearance rate of corticosterone of 13.8 ml min^{-1} kg^{-1}. In view of the high metabolic clearance rate of the hormone, the low amount of the hormone available within the gland cannot maintain a normal plasma-corticosterone level for more than some minutes, a feature that is in agreement with the mammalian pattern. Further investigations in the duck have shown that exogenous corticosterone (tritiated) is very rapidly

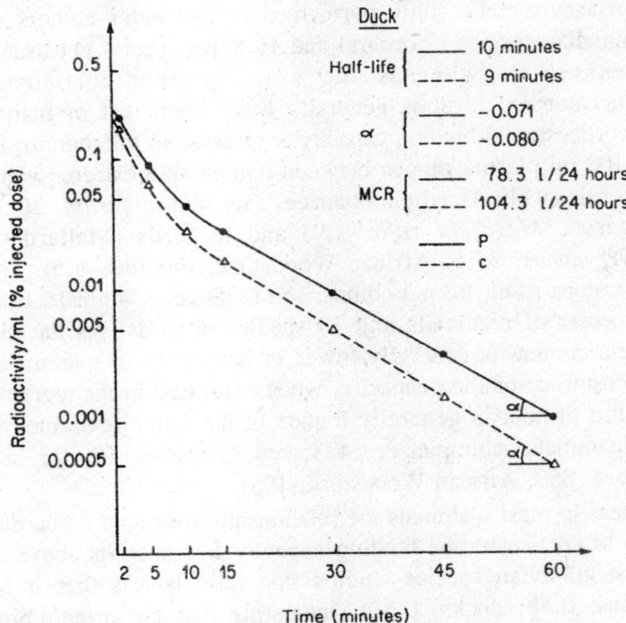

Fig. 3.2. Biological half-life of [^{3}H] corticosterone in the duck, and therefrom the calculated metabolic clearance rate (MCR).

metabolized, and several labeled metabolites appear within the blood stream, in increasing number and concentration with time, e.g., 30 minutes after a single intravenous injection, only 52% of the plasma radioactivity still belongs to corticosterone. Among the newly formed metabolites, 11-dehydrocorticosterone could be identified. When measurements of disappearance rate of intravenously injected labeled corticosterone are used as an index to the peripheral metabolism of the hormone, chromatographic purification of each plasma sample is required.

As has first been shown in mammals, plasma corticosteroids are strongly bound by a specific binding protein, corticosteroid-binding globulin (CBG), or transcortin. Independently of transcortin, which exhibits high affinity but low capacity for corticosteroids, and which has the major role in binding of corticosteroids, albumin constitutes a second corticosteroid-binding system in the plasma, characterized by a low affinity together with a high capacity. On the other hand, it has been shown that transcortin-bound corticosteroids are biologically inert, while the free, nonprotein-bound forms are the immediately active forms.

1. From a systematic study performed by the latter authors on 28 mammalian species (7 orders) and 16 avian species (11 orders), it appears that, whereas the avian plasma corticosteroid (corticosterone) level is generally lower than that of mammals, the corticosteroid-binding capacity, expressed in micrograms bound per 100 ml plasma ranges between comparable extreme values in mammals (white-faced chimpanzee, *Pan schweinfurthii*, 30.5; red kangaroo, *Megaleia rufa*, 1.9) and in birds (Mallard, *Anas platyrhynchos*, 35.5; African Woodstork, *Ibis ibis*, 6.5). Similar conclusions result from a comparison in domestic animals, between 55 species of mammals and 18 species of birds. As the plasma protein content is generally lower in birds than in mammals, the corticosterone-binding capacity, when expressed in micrograms per 100 gm protein is generally higher in the latter (extreme values for mammals—chimpanzee, 433; red kangaroo, 19; for birds—Mallard, 859; African Woodstork, 196).
2. Whereas in most mammals the relationship of plasma corticosteroid level to corticosterone-binding capacity lies near or above 1, in almost all avian species studied this ratio is less than 1 (e.g., pigeons, 0.58; ducks, 0.30), indicating that the protein-binding capacity of transcortin exceeds markedly the endogeneous corticosteroid level.

3. Sex differences are apparent in the domestic fowl, in which roosters are found to bind 4 μg cortisol per 100 ml plasma, compared with 8 μg% in nonlaying hens and 12 μg% in laying hens. Estrogen would then appear to have stimulating action on transcortin synthesis in birds similar to that in mammals. On the other hand, no sex differences have been found in the species studied by Steeno and De Moor (1966).
4. Intraspecies differences in corticosterone-binding capacity have also been found among ducks (e.g., 35.5 μg per 100 ml in Mallards; 33.9 in Indian runners; and 20.5 in other domestic ducks.

Functions of adrenocortical hormones

In birds, as in other vertebrates, adrenal function is vital. Adrenalectomized ducks and chickens usually die within 6-60 hours, unless given replacement therapy. Although less intensive investigations have been performed on the various actions of adrenocortical hormones in birds than in mammals, there is good evidence that the avian corticosteroids have the same major metabolic roles (e.g., on carbohydrate and electrolyte metabolism) as in mammals.

Carbohydrate metabolism

Riddle (1937) was the first to show clearly the hyperglycemic effect of adrenal extracts on hypophysectomized, thyroidectomized, or partially adrenalectomized pigeons. Later, several authors have demonstrated in the domestic fowl that corticosterone, and also hydrocortisone (but not cortisone), induce marked hyperglycemia, together with enhanced accumulation of glycogen in the liver. That increased liver glycogen is due to stimulation of hepatic glycogenesis rather than to impaired utilization of blood sugar is proved by the fact that corticosterone administration does not affect the *in vivo* oxidation of [^{14}C] glucose to CO_2. On the other hand, the modifications of carbohydrate metabolism are closely correlated with increased nitrogen output, indicating a stimulation of protein catabolism (gluconeogenesis). Thus, except for the relative ineffectiveness of cortisone in birds, the glucocorticoids behave much as in mammals. Finally, it must be stated that, despite the still questionable participation of the peculiar avian sacral "glycogen body" in carbohydrate metabolism, corticosterone is one of the few hormones that has been found to be active in glycogen enrichment of this organ in chickens.

Lipid metabolism

Corticosterone and hydrocortisone (but not cortisone) induce lipogenesis in the fowl, as indicated by elevation of plasma lipids, increased carcass (mainly subcutaneous) and visceral fat deposition, and generally enhanced liver fat. From histochemical studies it appears that when increased deposition of hepatic lipid occurs, the fat accumulates around the blood vessels, without any evidence of degenerative changes in liver cells. Corticosterone-induced lipogenesis does not result from changes in rate of fatty-acid metabolism, as no difference was evident in either the amount of *in vivo* oxidation of [^{14}C] palmitic acid, or in the rate of related $^{14}CO_2$ exhalation. On the other hand, injected [^{14}C]-glucose was directed mainly into lipid metabolism. As the same lipogenetic action of corticosterone treatment occurs in hypophysectomized chickens, this effect cannot be related to inhibition of some hypophysial lipolytic factor (ACTH, prolactin). *In vitro* studies on 3-4-week-old ducklings have also shown that, up to very high concentrations (10 μg per milliliter of medium), corticosterone has no direct effect on adipose tissue.

Protein metabolism—Growth

Despite a significant hyperphagia, corticosterone- or hydrocortisone-treated chickens exhibit marked inhibition of growth. The stimulating action of corticosteroids on nitrogen output, has already been mentioned, and Nagra and Meyer (1963) found in growing chickens that after corticosterone treatment there was a reduction in carcass protein and a decreased rate of transformation of injected [^{14}C] glucose into proteins.

Electrolyte metabolism

Kidney. As in mammals, the avian kidney is one of the major target organs of the corticosteroids, aldosterone appearing in birds also to be very potent in causing sodium retention. According to Brown et al. (1958a), in chickens with ureters surgically exteriorized, both cortexone acetate and cortisone increased urine flow, but deoxycorticosterone acetate decreased markedly the excretion of sodium and of potassium. In salt-loaded ducks, aldosterone lowers markedly the volume and sodium concentration of urine. Adrenalectomy in the duck causes increased sodium loss in the urine, while large doses of aldosterone or corticosterone reduce sodium excretion.

Salt glands. Recent investigations have demonstrated that adrenocortical hormones are also involved in the regulation of extrarenal electrolyte excretion in birds with active nasal glands. The nasal (supraorbital) gland of several saltwater species is known to be able to secrete a fluid, mainly a sodium chloride solution, that is hypertonic to the plasma. Due to its concentrating power, which is higher than that of the kidney, the nasal salt gland permits the animal to remain in water balance, even for prolonged periods of high salt diet. Except for the Falconiformes, no similar function has been described for the nasal glands of terrestrial birds, although the microscopic anatomy of the gland of the latter is very similar to that of aquatic birds. The size of nasal gland, the length of the secretory tubules, as well as the salt-concentrating capability of the glands from a number of marine species were found to be closely correlated with the ecology of the species. Several investigations on ducks and gulls have placed emphasis on the controlling role of the adrenal cortex on nasal-gland secretion. Acute salt load is followed by a biphasic reaction, with an initial diuresis followed by increased output of nasal secretion. The extrarenal response is increased by administration of adrenocorticosteroids or ACTH. On the other hand, hypertonic saline loading increases synthesis and release of corticosterone, which compensates for a correlative enlargement of the extracellular fluid volume. In turn, secretion by the nasal gland is significantly decreased by adenohypophysectomy and is suppressed by adrenalectomy. However, treatment of adenohypophysectomized ducks with ACTH restored the plasma corticosterone concentration, together with the extrarenal excretion of water and electrolytes, to normal. Finally, corticosterone has been found more effective in salt-gland stimulation than cortisol or cortexone, whereas neither aldosterone nor neurohypophyseal hormones seem necessary for the normal function of the nasal glands. The major glucocorticoid component of avian adrenocortical secretion seems thus to display a preeminent part of the controlling mechanism of extrarenal electrolyte excretion in birds. More recently, prolactin has also been shown to be capable of increasing nasal salt-gland secretion in ducks.

Reproduction

In view of the marked metabolic effects of corticosteroids, it is not surprising that adrenalectomy or corticosteroid administration have been reported to interfere with reproduction. Adrenalectomy has been claimed to induce pronounced atrophy of the testes and to inhibit the development of the right gonad in ovariectomized chickens. On the other hand, administration of cortisone acetate was followed by testicular atrophy, and corticosterone injections by depression of egg laying in chickens; whereas others have reported an androgenic effect of deoxycorticosterone acetate on comb size or a stimulating effect on testes. Induced states of hyper- and hypoadrenocorticalism elicited pathomorphic changes in reproductive systems of the pigeon.

Regulation of adrenal cortical function

Hypothalamic-hypophyseal control

Pituitary control. Although a number of experiments with mammalian ACTH or avian pituitary extracts have suggested the occurrence of a classic pituitary-adrenocorticotropic control scheme in birds as in other vertebrates, this control may differ in some respects from that of mammals, as the level of adrenalcortical function of hypophysectomized birds differs from that in hypophysectomized mammals. Whereas in the latter, hypophysectomy leads to a drastic atrophy of the adrenals, and an associated fall of 70-80% in plasma corticosterone, reveal that, in hypophysectomized birds, adrenal atrophy is variable and, moreover, that plasma-corticosterone levels remain at about 40-60% of controls. The adrenal corticosterone content appears more affected, whereas the aldosterone level remains unaltered. One may speculate that the high residual plasma-corticosterone level in hypophysectomized birds, as compared with that of hypophysectomized mammals, maybe due to a lowered metabolic clearance rate. In fact, hypophysectomized ducks exhibit an 83% increment of the half-life of corticosterone, and a correlative 42% reduction of metabolic clearance rate. However in hypophysectomized rats, the half-life of corticosterone also exhibits a 50% increase, so that the obvious discrepancy in plasma

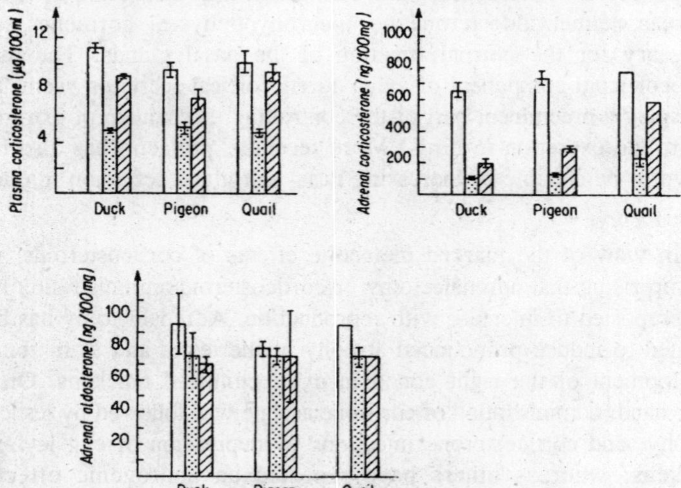

Fig. 3.3. Plasma and adrenal corticosterone and aldosterone in control (clear), hypophysectomized (dotted), and pituitary autografted (hatched) ducks, pigeons, and quail.

corticosterone in either birds or mammals may more likely be due to higher residual corticosterone secretion rather than to some quantitative difference in the peripheral metabolism of the hormone.

Two theories have been proposed to account for these differences: (1) the occurrence of an autonomous adrenal cortex secretory activity, independent of pituitary ACTH, and (2) the possible action of an extrahypophyseal corticotropic substance, either ACTH or an ACTH analogue. The latter hypothesis receives support from the fact that in both the domestic fowl and pigeon an ACTH-like activity has been demonstrated in the median eminence of the hypothalamus. Peczely et al. (1970) have shown that this ACTH-like activity in the median eminence of the pigeon, 1 month after hypophysectomy, is still as high as in control animals. Furthermore, hypophysectomized pigeons respond to formalin stress by morphological activation of cortical tissue, while short-term surgical stress increased plasma corticosteroid levels in hypophysectomized chickens. On the other hand, Frankel et al. (1967a) have recorded in the hypophysectomized fowl a drop in plasma corticosteroid to nearly zero after injection of dexamethasone. It seems possible, therefore, that in birds, a production of ACTH or some related substance may persist after hypophysectomy and may be responsible to some extent for the high residual plasma corticosteroid values observed and for adrenal stimulation in the absence of the pituitary.

Hypothalamic control. In the hypothalamic control of ACTH release, there is a further discrepancy between birds and mammals. In mammals, numerous investigations have shown that pituitary autografts are ineffective in preventing a drastic fall in plasma corticosteroids to hypophysectomy levels. In birds, on the other hand, although pituitary autografting is often accompanied by morphological atrophy. The basal plasma corticosteroid level, in long-term autografted ducks, pigeons, and Japanese Quail, lies far above the hypophysectomy level. Similarly, the adrenal corticosterone level is significantly raised, and the half-life and metabolic clearance rate for corticosterone is restored to normal values. Moreover, the ACTH content of pituitary transplants, in autografted pigeons, was found to equal that of the pituitary *in situ*. This is in contrast to the observations of Resko et al. (1964), who reported that, in the fowl, autografts have no effect on plasma corticosteroid levels in the adrenal effluent blood. This may reflect a species peculiarity. However, the postoperative time lapse seems very important, as in ducks the plasma corticosteroid level decreases equally within the first 3 weeks after either autograft or hypophysectomy, but

2 months later the plasma corticosterone level is elevated to the typical 80% of controls in autografted animals. Furthermore, the reestablishment of plasma corticosteroid levels after autograft is concomitant with the recovery phase of the cytofunctional characteristics of the grafted tissue in this species. Thus, the above-mentioned apparent inability of pituitary grafts to promote stimulation of adrenal cortical function in chickens may be due to an incomplete recovery of the grafted tissue, since measurements were not recorded beyond 40 days postoperatively. In fact, corticotropin-releasing factor (CRF) similar to that formed in mammals, has been reported to occur in median eminence extracts of pigeons, hens, and ducks. In the pigeon, hypophysectomy leads to the disappearance of the CRF activity in the median eminence extracts, suggesting a rapid depletion of CRF into the general circulation, while in autografted pigeons, CRF activity in the median eminence is retained. The possible high release of CRF into the general blood stream by hypophysectomized or autografted animals may account for the maintenance of some ACTH release by the ectopic pituitary. As a matter of fact the destruction of the mediolateral posterior area of the hypothalamus in pituitary-autografted pigeons prevents the recovery of the plasma corticosterone concentrations to the near normal levels that are progressively attained in the simply autografted controls. On the other hand, it has been emphasized that destruction of the median eminence in addition to autografting in the duck was without effect on the subnormal plasma corticosteroid level, which is usual in animals with pituitary autografts. The site of origin of the CRF is still unknown. However, Frankel et al. (1967b) have noticed marked depression of corticosterone secretion in the fowl after lesions within the ventral hypothalamus, whereas similar results were obtained in the pigeon after destruction of a more dorsal area of the hypothalamus. Finally it must be emphasized that the neuroendocrine control of the adrenocortical function probably involves central catecholaminergic and serotoninergic neurons, since their pharmacological suppression, which can be traced within the median eminence, coincides with a 40 to 60% fall in the mean level of the plasma corticosterone.

Rhythmicity of adrenal cortical function

Diurnal rhythm. The occurrence of diurnal fluctuations in adrenal-cortical function, as measured by plasma or urinary corticosteroids, have been clearly demonstrated in mammals. In general, mammalian species, such as monkey and also man, that are diurnally active has more active adrenals in early morning than in late evening hours, the reverse occurring in species such as rats or mice that are predominantly nocturnal animals. Extensive investigations have been made in this respect in a typical diurnally active bird, the Japanese Quail, with simultaneous measurements of plasma and adrenal-corticosterone levels, together with autographic and body-temperature records. Quail reared in isothermic rooms but under natural light conditions (June), exhibit a marked daily rhythm for both adrenal-cortical parameters studied very similar to that of diurnal mammals, with a steep increase during the second half of the night, a peak toward the end of the night hours, and a progressive lowering during day hours. The same adrenal-cortical cycle has been observed in ducks. On the other hand, if quail are submitted to constant environmental conditions (constant light at 3 lux) an "endogenous" rhythm of plasma corticosterone and locomotor activity takes place, with a "circadian" periodicity of 22.50 hours. The cyclic evolution of the plasma and adrenal corticosterone has also been studied in quail submitted to various photoperiodic schedules, e.g., 12L:12D (130 lux), 12L:12D (3-5 lux); 6L:18D (130 lux, starting

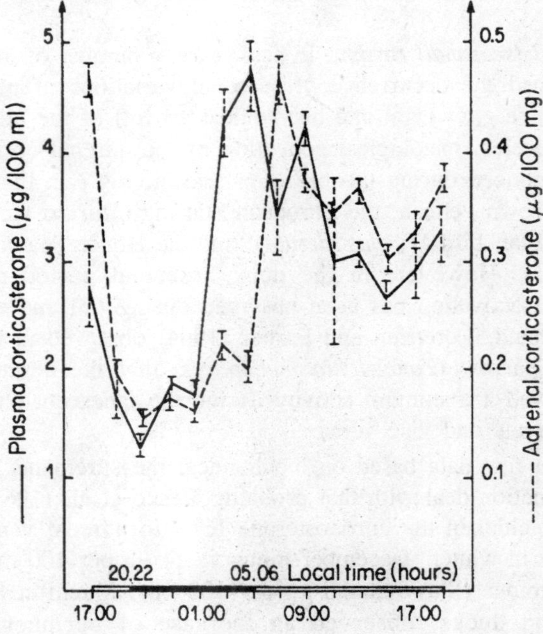

Fig. 3.4. Daily rhythm of adrenal function of Japanese Quail exposed to natural light conditions.

either at 7:00 AM or at 7:00 PM); 18L:6D (130 lux). The adrenal rhythms were always strictly synchronized by the photoperiod, the daily increment of corticosterone occurring always during the scotophase, followed by a progressive decrease during the photophase. The adrenal rhythm also remains in constant phase relationship with the light-dark-entrained rhythms of locomotor activity and of body temperature. Furthermore, the corticosterone rhythms, as well as the locomotor-activity rhythm, of quail remain in phase with the environmental light-darkness cycles, even when the birds are submitted to more sophisticated ahemeral photoperiods, such as 6L:7D (3 lux), 9L:26D (3 lux), or 26L:9D (3 lux), which range definitely below (13 hours), or beyond (35 hours), the classic circadian span (20-28 hours). That the diurnal rhythm of adrenocortical function is dependent on its complex neuroendocrine control machinery is shown by (a) the occurrence of a diurnal rhythm in the pituitary ACTH content, which precedes by a few hours, the phase of the plasma corticosterone rhythm, and (b) the suppression of a 24 hour rhythm of the plasma corticosterone by either autografting the pituitary onto the kidney, or blocking pharmacologically the catecholaminergic or serotoninergic neurons in the central nervous system.

Annual (seasonal) rhythm. In past years, a number of investigators have claimed the occurrence of seasonal variations of morphologic parameters (e.g., weight and histological aspect) of the bird adrenal. In most species, histological activation of the adrenal cortical tissue has been noticed during the breeding season, e.g., in the duck, the hen, the Brown Pelican, the European Starling (*Sturnus vulgaris*), the European Blackbird (*Turdus merula*), and the House Sparrow (*Passer domesticus*). However, in the duck a second period of adrenal histological activation has been observed during fall and winter. On the other hand, Lorenzen and Farner (1964) observed in the White-crowned Sparrow (*Zonotrichia leucophrys*) that the adrenal-cortical tissue showed a maximum activity during the quiescent phase of the testicular cycle and vice versa.

Only a few data based on biochemical measurements of adrenal cortical function deal with that problem. Resko et al. (1964) claimed that in the chicken the corticosterone level in adrenal venous blood was higher in winter (December-January: 24 μg per 100 ml plasma), than in summer (May-June: 6 μg per 100 ml), Machi et al. (1967), working on ducks, observed an increase of peripheral plasma corticosterone levels from February-March (21.05$\pm$2.53 μg per 100 ml) to April-May (33.57±1.86 μg per 100 ml). A systematic study on monthly fluctuations of peripheral plasma corticosterone levels has been performed on ducks reared outdoors under field conditions of the Mediterranean region and fed *ad libitum*. The plasma corticosterone revealed marked annual variations, with a minimum in March (11.83±1.61 μg per 100 ml) and a maximum in October-November (27.30±1.56 μg per 100 ml). During the summer months (May-September), the plasma corticosteroid level was at a plateau at about 17 μg per 100 ml.

The marked discrepancy among the cited results obviously may reflect species variations, as well as differences in climatic, nutritional, or other environmental factors to which the birds were submitted. Finally, the different techniques used make difficult a valid comparison of the available data. However, three types of correlations may be considered: (1) a positive correlation between the adrenal cortical cycle and the photoperiodically induced breeding season, (2) a negative correlation between the same parameters, (3) a positive correlation between cortical stimulation and low environmental temperatures. The problems raised herein will be discussed in the following sections.

Gonadoadrenal interrelationships

Although in a number of bird species the adrenal glands enlarge annually during the reproductive season, some species—e.g., the male White-crowned Sparrow and the male duck—show an inverse phase relation between the two endocrine cycles, which may point to a possible antagonism, at least between testes and adrenals. For instance, in ducks reared under field conditions, the extreme annual plasma corticosterone levels were 27.30 μg per 100 ml in November, as compared with 11.8 μg per 100 ml in March, while the extreme, plasma testosterone levels, measured by gas chromatography, were 4.6 ng per 10 ml in November as compared with 16.1 ng per 10 ml in March. Such an antagonism, in fact, has been shown experimentally.

Most of the scant data available on gonadoadrenal relationships in birds are related to the male, and the hitherto known effects of testosterone on avian adrenocortical function seem very close to those described in mammals. In cockerels and in White-crowned Sparrows castration leads to adrenal hypertrophy, and testosterone injections were found to prevent this effect according to Breneman (1941) and Kar (1947), but not according to Nagra et al. (1965). In fowl, castration also elevates the corticosterone level in adrenal venous blood, while testosterone injections lower the adrenal effluent corticosterone to control values. On the other hand, testosterone-injected gonadectomized fowls exhibited the same biological half-life of corticosterone as untreated castrates, when measured by the unlabeled corticosterone-loading method of Kitay (1961). As in several investigations on mammals, testosterone administration in intact androgen-secreting fowls had no effect on adrenal weight nor on the amount of corticosterone in adrenal effluent blood, but the same investigators showed that the biosynthesis of corticosterone by adrenal tissue *in vitro* was inhibited by the male hormone.

Further evidence for a strong androgen inhibition on corticosterone secretion without any impairment of aldosterone secretion in the duck. It may also be recalled here that castration in quail and ducks submitted to either long or to short days elicits significant activation of the ACTH cells within the anterior pituitary. On the other hand, male pigeons seem to behave in a different way than fowls and ducks, since castration as well as testosterone injections have been found to elicit adrenal atrophy, together with cytological and histochemical pictures of adrenal regression.

Estrogens have been less extensively investigated thus far. However, adrenal hypertrophy, which is classic in mammals, also occurs in estrogen-treated fowls, and pigeons, together with histochemical pictures of cortical stimulation. According to the latter, no significant adrenal effect was noticed in spayed pigeons. Zarrow and Baldini (1952) failed to induce adrenal hypertrophy in estrogen-treated Bobwhite Quail (*Colinus virginianus*). Progesterone, on the other hand, has been shown to induce increased phosphatase activity in the adrenal cortex of pigeons of both sexes.

Environmental (external) factors

Photoperiod. Despite the many observations of the adrenal-cortex activity as a function of the seasonal or experimental increase of day length, no decisive conclusion can be drawn as to whether or not a direct causal relationship exists, i.e., whether photoperiod acts as a specific source of information in the control of the adrenocorticotropic axis as for other functions (e.g., the gonadal function). Enhanced locomotor activity and nutritional functions, which are simultaneously activated throughout the seasonal long-day period could conceivably act as light-induced secondary stimulators for adrenal-cortical function. On the other hand, for those species, in which an inverse correlation exists between increasing light-dark ratios and adrenal-cortical activity, it has been seen that there are arguments to consider these correlations as a result of gonadoadrenocortical interactions.

The diurnal periodicity in adrenal cortical function probably provides a better model than do seasonal rhythms for the study of photoperiod on the function of the adrenal cortex. Whereas all of the above cited light-dark schedules entrained the plasma and adrenal corticosterone rhythms in the quail, the testes of the birds were either developed—e.g., in the natural summer environment, and in the schedules: 12L:12D (130 lux), 18L:6D (130 lux); 6L:7D (3 lux); 26L:9D (3 lux)—or regressed—6L:18D (130 lux); 12L:12D (3 lux); 9L:26D (3 lux). Because of the limiting action of either day length or intensity on the gonadotropic axis, the two neuroendocrine mechanisms were dissociated in these experiments, and the high degree and plasticity of light-dark dependence of the adrenal-cortical function appears here to be clearly independent of gonadal function. However, here again it is not possible to assess the direct light sensitivity of the adrenocorticotropic machinery since the effects of different light-dark ratios may really be derived from other direct photoperiodic regulations, such as in the control of locomotor activity and food intake, which have marked metabolic impacts, and which are entrainable by light-dark cycles.

Temperature. Hohn's description of a second peak of morphological stimulation of the duck's adrenal during the cold season points to an adrenal cortical adaptation to chronic cold environment. Knouff and Hartman (1951) also have reported on a histochemical activation of cortical tissue in Brown Pelicans exposed to cold. The annual adrenal rhythm in the duck, which has been analyzed above, does not show any definite correlation between environmental temperature and function of the adrenal cortex. However, the comparison between the plasma-corticosterone level of the animals studied in this experiment and held outdoors in November (26.40 μg per 100 ml) and ducks from the same flock that had been kept at the same period for 2 weeks indoors at +20°C (18.8 μg per 100 ml) reveals a possible stimulating action of chronic exposure to cold. On the other hand, Boissin (1967) has studied the adrenal repercussion of *acute exposure to cold* in the duck. When measured at 4, 12, and 24 hours after cold exposure (from +25°C to +4°C), plasma corticosterone was significantly increased respectively by 25, 40, and 21%, showing a rapid but transient response of the adrenal cortex. Hypophysectomized ducks and specimens bearing pituitary autografts were submitted to the same treatment. The hypophysectomized ducks, which had plasma corticosteroid levels at 47% of intact controls, failed to exhibit adrenal reaction to cold, whereas the autografted birds had a positive reaction, though shorter than the intact controls. The 25°C autografted ducks had plasma corticosterone levels at 76% of controls. After a 4-hour exposure to cold the plasma corticosteroids increased by 28%, after 12 hours by 48%. However, 24 hours later the plasma corticosteroid values had returned to normal levels for birds with pituitary autografts.

Feeding. Although there has been much interest in the gonadal effects of underfeeding, few studies have dealt with possible adrenal involvement. Nevertheless, several species, when fed a restricted diet, show some evidence of adrenal-cortical stimulation. Underfed pigeons and chickens showed adrenal hypertrophy, while Hartman el al. (1954) noted a higher rate of mitoses associated with lipid depletion in the adrenal cortex of undernourished Brown Pelicans. In the turkey, starvation leads to a significant increase in plasma corticosterone. Enhanced plasma corticosterone levels also were observed in male ducks starved for 17 days, and the absence of a significant concomitant alteration of the biological half-life of corticosterone in those animals points to a definite stimulation of the adrenal-cortical function.

Medulla

Hormones

As in other vertebrates, the medullary cells secrete catecholamines. The avian medulla has first been considered as secreting mainly norepinephrine, as the predominance of that hormone has been evidenced by histochemical methods in the pigeon and in the Gentoo Penguin, and by chromatographic procedures in the fowl (70-80% of the total catecholamine content of the gland). On the other hand, adrenaline-containing granules and noradrenaline-carrying granules have been isolated from homogenates of chicken chromaffin tissue by centrifugation in a sucrose-density gradient. Systematic investigations using histochemical assays have shown that there are marked species differences in the norepinephrine-epinephrine ratio; (1) medullas of Pelecaniformes, Ciconiiformes, and Galliformes are essentially norepinephrine producers; (2) Passeriformes are mainly epinephrine secretors; and (3) Columbiformes and Cuculiformes have both hormones present in their adrenals, with a predominance of norepinephrine in the former and of epinephrine in the latter. According to Ghosh, this could reflect an evolutionary trend, the hormonal methylation being mainly absent in more primitive avian orders. On the other hand, the higher levels of epinephrine in comparison with norepinephrine in peripheral blood of the chicken, turkey, and pigeon have been interpreted as a reflection of a higher release of epinephrine by the adrenals of those species, an inverse ratio existing in the duck blood. However, the neuronal source of both catecholamines might be partly involved in the plasma levels of the hormones, and the question hence needs further clarification. On the other hand, the concentration of norepinephrine has been found to be significantly higher in plasma of female than male chickens, but there was no sex difference in concentrations of epinephrine.

Regarding the control of epinephrine biosynthesis, a cytochemically assessed rise of epinephrine and a concomitant depletion of norepinephrine in cortisone-treated pigeon adrenal medullas indicates the possibility of conversion of the latter to the former by a methylation mechanism similar to that which exists in mammals.

Role of the adrenal medulla

Although some effects of epinephrine and norepinephrine in birds have been reported, little is known about the precise physiological role of the avian adrenal medulla, as other (nervous) sources actively participate to the production of catecholamines.

Heat production

Norepinephrine may have a thermogenic function in mammals with brown adipose tissue, whereas in species without brown fat it does not. The hitherto studied birds behave like the latter mammals. In the young chick, neither noradrenaline nor adrenaline arrest the fall of body temperature induced by cold and anesthesia. Indeed, noradrenaline lowers the rectal temperature of anesthetized chicks exposed to cold, and the same hypothermic effect has also been shown in quail 6-9 weeks old. The physiological significance of the 60% release of the norepinephrine stored in the adrenal medulla of pigeons exposed to acute cold therefore remains obscure.

Lipid metabolism

In the neonatal fowl noradrenaline (300 μg/kg) induces a rise in plasma free fatty acids. This response increases from hatching to 2 weeks of age. By 8 weeks, the response is much reduced. In adult fowls, in ducklings, and in adult ducks and turkeys, neither epinephrine nor norepinephrine, when used at physiological doses, either *in vivo* or *in vitro*, bring about release of free fatty acids from adipose tissue, nor are the catecholamines able to cause any significant change of liver triglycerides in the duck. However, epinephrine stimulates lipolysis in pigeon adipose tissue *in vitro* and norepinephrine induces a steep increase in plasma free fatty acids in 6-9-week-old quail. In geese, both catecholamines, whether injected or infused, cause an elevation of plasma free fatty acids, although the response is considerably smaller than that produced by glucagon.

As norepinephrine is not a thermogenic factor, but induces hypothermia in birds exposed to cold, it seems unlikely that it is directly responsible for the increase in plasma free fatty acids that results from cold exposure.

Carbohydrate metabolism

The hyperglycemic effect of epinephrine has been known for years in the duck, in the chicken, in the pigeon, and in the goose. In the duck, infusion of norepinephrine causes also a significant hyperglycemia, while the hormone produces only minimal changes in geese and turkeys. In pigeons, norepinephrine is practically as potent as epinephrine with regard to hyperglycemia. In the neonatal fowl, norepinephrine (300 μg/kg) increases plasma glucose, but no hyperglycemia developed in 1-2-week-old chicks. A significant hyperglycemia again occurs in 4-8-week-old chickens. On the other hand, in 6-9-week-old Japanese Quail, there is an unexpected effect of norepinephrine—a significant fall in plasma glucose within 30 minutes after epinephrine injection has been observed. Finally, epinephrine in large doses also has been found to enhance significantly the glycogen content of the sacral glycogen body in chickens, whereas the liver glycogen was decreased.

Blood pressure

Perfusion of anesthetized chickens with norepinephrine and even more so with epinephrine, leads to significant rise in the systolic and diastolic blood pressure. From histochemical studies Ghosh (1973) concluded that both catecholamines are equally effective as pressor hormones in the House Crow (*Corvus splendens*) and the fowl. Heart rate is unaltered by norepinephrine or even decreased with epinephrine unless the vagi are cut; in the latter case both catecholamines increase heart rate slightly.

THYROID GLAND

Morphology

The avian thyroid gland is paired, and its two dark red, ovoid lobes are located low in the neck, internal to the jugular vein, and external to the carotid (at its junction with the subclavian artery) and to the trachea. Structurally the avian thyroid gland is composed, as in other vertebrates, of roughly spherical follicles, each consisting of a colloid-containing lumen surrounded by a single layer of cuboidal epithelial cells. An extensive vascular supply is provided to the follicles.

As in all vertebrates, the avian thyroid gland arises as a midventral outpocketing from the floor of the pharynx, which forms the longitudinal mesobranchial groove. Later on, the mesobranchial groove disappears anteriorly, leaving as its only remnant an epithelial thickening, which is the first visible primordium of the thyroid gland. The original primordium soon becomes bilobed and gradually migrates back as paired organs toward its adult location. Progressively invading mesenchyme and blood vessels crowd the epithelial mass of each lobe into solid cylindrical cords. Midway in the incubation period the cords begin to break up in small cell groups that acquire central lumina as a result of secretion and coalescence of colloid droplets originating in the follicular cells. From then on, the thyroid follicles are fully differentiated, and their functional state already is evidenced by a more or less strong radioiodine (^{131}I) uptake. During the latter half of incubation, the main feature consists in follicular growth in size and number.

Normal and TSH-stimulated thyroid glands of the fowl have been investigated by Fujita (1963) with the electron microscope. Two types of intracellular droplets, respectively of high and of low electron densities, were identified. Both are enclosed by a limiting membrane composed of two dense and a less dense layer, similar to that of the other cytoplasmic membrane systems. The dense granules appear to be formed in the Golgi field, while the less dense ones appear to derive from the endoplasmic reticulum. The latter droplets are considered to contain colloid, as they increase markedly in number and size, together with endoplasmic sacs, after TSH stimulation, and several droplets seem to be secreted into the follicular lumen through an opening occurring at a junction between the apical plasma membrane and the limiting membrane of the droplet. Some cellular processes containing numerous low-density droplets and rough-surfaced endoplasmic sacs are also observed after TSH administration. On the other hand, after TSH treatment, a distinct endoplasmic reticulum and large droplets of low density may be seen close to the basal plasma membrane, suggesting a direct secretion of the content of droplets and endoplasmic sacs into the pericapillary or interfollicular space from the basal part of the follicular cell. Finally, a central flagellum with a basal corpuscle is observed in some follicular cells of young and adult fowls. Its function has not been clarified.

Hormones and Metabolism

There is an extensive literature in recent years on thyroid function in birds, especially since the development of radioisotopic methods. From these data, it appears that the hormones of avian thyroid and the sequences of their synthesis, metabolism, and catabolism are fundamentally similar to those of mammals despite some peculiar features, which generally concern quantitative rather than qualitative aspects of metabolism. It is well established that the biosynthesis of thyroid hormones occurs within the polypeptide chain of the specific protein thyroglobulin. As in mammals, the major thyroid iodoprotein present in avian thyroid follicles has a sedimentation coefficient close to 19 S. In the duck and chick, the 19 S thyroglobulin concentration among soluble thyroid iodoproteins is about 94% (it is 92% in man, 93% in rats, and 94% in dogs). Other minor thyroid iodoproteins have also been identified in birds as in mammals—e.g., a 27 S thyroglobulin, in a concentration of 5% in ducks and 6% in chickens (8% in man, 7% in rats, and 6% in dogs), together with traces of 12 S thyroglobulin (a compound that has not been detected in man or dog and is present in traces in the rat.

Using radioisotopes, most investigators have identified the two classical thyroid hormones within thyroid hydrolysates, i.e., triiodothyronine and thyroxine, the latter being largely predominant—e.g., in the domestic fowl, in the Japanese Quail, in the duck, in the pigeon, in the Golden Bishop (*Euplectes afer*), and in the White-throated Sparrow (*Zonotrichia albicollis*). Similarly the two hormones were shown to occur in the plasma of the fowl, the Japanese Quail, the turkey and the duck. However only thyroxine has been identified by others in the thyroid gland of the domestic fowl. Whether the discrepancies noted in domestic fowl may originate from technical differences, from iodine content in the diet, or from variations in the strains studied needs further clarification.

From the numerous reports on avian thyroid function, and from several systematic comparative studies performed in birds and mammals an avian pattern of thyroid function may be outlined.

1. At the thyroid-gland level, the bird is characterized, comparatively with the mammal by (1) a higher total thyroid ^{127}I content; associated with (2) a higher thyroid T4 ^{127}I concentration, since the distribution of ^{127}I among iodinated thyroid compounds as well as the equilibrium of thyroid ^{131}I with ^{127}I in chickens is similar to that in rats fed a similar iodine diet; (3) an elevated net rate

of trapping of ^{127}I, as calculated from specific activities of serum iodide and thyroid ^{131}I uptake at early times after injection of radioiodide; and (4) a prominent iodine storage, leading to a prolonged thyroid retention of ^{131}I and to an almost flat thyroid ^{131}I output curve, both definite characteristics of birds.

2. At the peripheral-plasma level, avian thyroid metabolism exhibits further marked peculiarities: (1) In the absence of a thyroxine-binding globulin (TBG) in birds, the only available binding proteins lie within the prealbumin (TBPA) and albumin (TBA) zone. Chicken and pigeon sera bind both T4-^{125}I and T3-^{125}I to TBA, but among all vertebrates hitherto studied, the pigeon is the only species that binds T3 to prealbumin. On the other hand, although Tata and Shellabarger (1959) had claimed that both T3 and T4 were equally bound to plasma albumin in chickens, others have pointed to a greater affinity to T4 than T3 for the plasma protein binding sites. These peculiarities of serum-binding proteins for thyroid hormones may account for some further characteristics, such as (2) lowered amounts for either total serum iodine, or PB ^{127}I. (3) While avian levels of total thyroxine range between classic values found in other warm-blooded vertebrates, the free plasma thyroxine, which might be considered a more accurate index for thyroid status than total T4, is relatively high in birds in comparison with most mammals. (4) As a consequence of the foregoing, the peripheral metabolism of both thyroid hormones is remarkably rapid in birds, as evidenced by their short biological half-lives. Since the clearance rate of labeled iodide from the plasma appears equally slow in birds and in rats with perchlorate-blocked thyroids, the fast peripheral metabolism of the thyroid hormones in birds (almost four times as rapid as in rats) may account for the marked difference in the half-life of thyroxine-I, if measured either from total plasma samples, or from only the protein-bound I fraction of the plasma, that is usual in birds but not in rats. (5) On the other hand, emphasis has also been laid recently on the fact that, unlike in rats, the separation by Sephadex gel chromatography of the ^{125}I-labeled compounds in sera of several bird species (fowl, duck, pigeon, Japanese Quail) points to a large amount of nonhormonal iodoproteins within the avian PBI. This parameter is therefore a rather inaccurate index of thyroid hormone concentrations in the blood of birds.

Very little is known as yet about the role of storage and discharge of thyroid hormones from extrathyroid tissues in the regulation of the

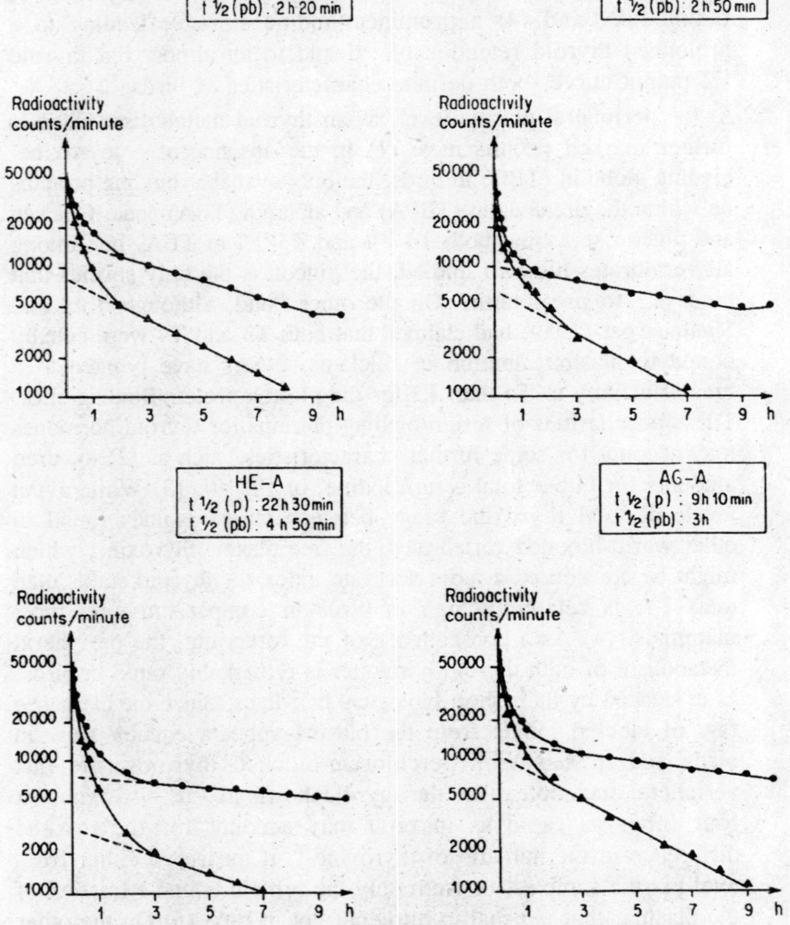

Fig. 3.5. Biological half-life (t $^1/_2$) of [^{131}I] thyroxine in the duck. C=controls; HE=hypophysectomy; AC=pituitary autograft; A=iodine-adequate diet; L=low-iodine diet.

effective level of circulating thyroid hormones in birds. On the other hand, it has been well established from studies performed on the domestic fowl with either artificial anus or canulated bile ducts and ureters that, as in mammals, the main route of excretion for both thyroid hormones is via the bile and feces, the urine containing almost exclusively iodine in inorganic form. Furthermore, chromatographic and electrophoretic analyses of bile samples from chickens injected with T4-^{131}I or T3-^{131}I have revealed that the assortment of ^{131}I-labeled metabolites identified in the bile, and hence the peripheral metabolism of both hormones, is similar to that known in mammals. Metabolism was found to occur by conjugation (mainly glucurono binding, and to a lesser extent, sulfate binding), deamination, and deiodination (with formation of diiodothyronine).

Similarly, extensive studies have shown that goitrogenic drugs have the same effect in birds as in mammals, both at the thyroid and at the peripheral level. Thiouracil, for instance, has marked extrathyroid effects in chickens as in rats, in reducing the peripheral utilization of thyroxine, while neither methimazole nor potassium chlorate have such effects.

Finally, the total thyroid activity, as measured by daily thyroid-hormone secretion rate ranges within limits close to those found in mammals. Unfortunately, data published until recently lack information on the iodine content of the diet of the animals studied.

Now very thorough investigations on the domestic fowl by Rosenberg et al. (1963, 1964) afford evidence of the definite importance of the alimentary iodine. The comparison of chickens fed an iodine-deficient diet (0.14 μg I^- per gram) with others receiving an iodine-supplemented diet (+2 μg I^- per gram) lead to the following conclusions: (1) The deficient chickens had consistently lowered (one-third) amounts of total ^{127}I in their thyroids. Despite an increased fraction of [^{127}I] thyroxine in the thyroid glands of the deficient chickens, the total intraglandular amount of the hormone was low in these birds. (2) The deficient chickens exhibited enhanced thyroid ^{131}I uptake (52% at 24 hours compared with 12% for the supplemented group). (3) With the deficient diet, ^{131}I incorporated among iodinated compounds of the thyroid gland reached an apparent equilibrium distribution with the preexisting stored ^{127}I compounds at 48 hours, while equilibration required approximately 4 days with the supplemented diet. (4) Within the plasma, the deficient diet lead to a marked fall in the total ^{127}I content (20% of the supplemented diet), as well as in the protein bound ^{127}I (60% of the supplemented diet). Similar results have been described in the duck.

Using an open two-compartment model devised by Wollman and Reed (1959), an attempt has been made to characterize the transport of radioiodide between blood and thyroid gland in chicken, in relation to iodide levels of the diet. This method demonstrated that the thyroid/serum radioiodide ratio after a single ^{131}I injection and the thyroid clearance constant for radioiodide varied inversely with the concentration of alimentary iodide.

All these results are of primary importance, as they clearly reveal that no valid assumption on possible intra- and interspecific differences in thyroid metabolism among the various bird species hitherto studied may be assessed without a precise statement on the dietary iodine intake of the birds.

Role of Thyroid Hormones

In mammals, triiodothyronine has been shown to possess 2-6 tinges the potency of an equimolar quantity of thyroxine as measured by a number of physiological tests. In birds, controversial results have been obtained. According to Gilliland and Strudwick (1953), triiodothyronine has more potency than thyroxine in blocking TSH release, whereas thyroxine was found more potent than triiodothyronine in preventing goiter in thiouracil-treated birds and in promoting oxygen uptake by myocardium in ducks. On the other hand, both hormones were claimed to be equipotent in preventing goiter, in stimulating heart rate, and in counteracting the effects of propylthiouracil or radio-thyroidectomy on body and comb growth rates and on liver glycogen. A possible explanation for the different relative potencies of T4 and T3 could lie, as in mammals, in different binding affinities of the hormones for their carrier plasma proteins. But, as was mentioned previously, there is also some discrepancy among the available data concerning this point, so that further clarification is still needed.

Growth

As in other vertebrates, thyroid hormones are essential for regulating harmonious growth. Thyroidectomy in ducks and in domestic fowl caused marked retardation in growth, leading to the classic hypothyroid shortleg dwarfism, associated with abnormal fat deposition and sometimes obesity. Similar results are produced by radio-thyroidectomy, while triiodothyronine or thyroxine restore the body weight of the operated chicks. Goitrogen feeding yields less constant results. However, the growth rate of 3-week-old chickens was significantly retarded by administration of tapazole, and thyroxine (2-3 μg per 100 gm/day) counteracted completely the effect of the goitrogen. Similar results have been obtained by Raheja and Snedecor (1970) using propylthiouracil and triiodothyronine or thyroxine (0.3 μg per 100 gm) as replacement therapy.

On the other hand, thyroid-hormone administration to intact birds results only in moderate acceleration, if any, of growth. Singh et al.,

(1968b) noted improved growth rate in 3-week-old chickens injected with thyroxine (1-4 μg per 100 gm/day). However, higher doses (6 μg per 100 gm/day) accelerated catabolic processes and depressed the growth rate of the birds.

Heat production

In adult birds, the main role of the thyroid seems to be the maintenance of a generally normal metabolic rate. Although thyroidectomy is not lethal, the removal of thyroid hormones leads to a marked depression of heat production in the chicken, the pigeon, and the goose. Similar depression of the metabolic rate is obtained by chemical (goitrogen) hypothyroidism, while administration of thyroprotein enhances heat production in the pigeon, and in the fowl. Singh et al. (1968b) have observed that single injections of thyroxine and triiodothyronine or of a combination of the two produced only a small and transient (2-hour) rise in metabolic rate of chickens. The metabolic rate was depressed 24 hours after injection.

In the neonatal chick, both thyroxine and triiodothyronine, when injected intraperitoneally (300 μg/kg) were thermogenic. Rectal temperature increased significantly within 30 minutes when the chicks were maintained in a thermally neutral environment. Both hormones were equally efficacious in delaying the fall in rectal temperature when the chicks were exposed to cold. The stimulating effect of thyroid hormones on tissue metabolism has been well demonstrated in an older experiment by Haarmann (1936). Oxygen consumptions by heart, liver, and breast-muscle slices from chickens and pigeons were elevated up to 30% when incubated with minimal concentrations (10^{-10}–10^{-18}) of thyroxine.

Carbohydrate metabolism

Riddle and Opdyke (1947) found that in pigeons thyroxine reduces liver-glycogen level and produces mild hyperglycemia. Inversely, thyroidectomy causes hypoglycemia in the pigeon and in the duck. Radiothyroidectomy or feeding propylthiouracil to chicks induces hepatic hypertrophy and enhanced liver-glycogen levels, while injections of triiodothyronine or thyroxine prevent all of these effects.

Skin and feathers; molting

The thyroid gland exerts marked influence on the skin and its derivatives. The morphogenetic effect of thyroxine on the epidermis is already apparent in the chick embryo. At 14 days of incubation, the skin epithelium of untreated embryos is restricted to three undifferentiated cell layers, whereas embryos treated with 20 μg thyroxine showed seven cell layers with distinct germinative and corneal layers, a picture similar to that of 16-day-old untreated controls.

In the adult bird, the importance of thyroid hormone on plumage has been amply verified. Thyroidectomy- or goitrogen-induced hypothyroidism affects feather morphology markedly; the feathers become elongated, thinner, and lose their regular, rounded shape, acquiring an almost completely irregularly fringed contour, due to impaired development of the barbules. In some species, e.g., the domestic fowl, most melanin pigments disappear, whereas in others, thyroid glands have no marked effects on the feather pigments. These alterations are reversed by injections of thyroxine. Moreover, Kraetzig (1937), Svetsarov and Streich (1940), and Thapliyal et al. (1968) have shown a direct stimulating action of thyroxine on the feather follicle, where numerous mitoses are observable from the first day of treatment.

These classic observations have lead to the assignment to the thyroid gland of a major role in the onset of molt. From the considerable literature published on this topic several conclusions may be drawn: (1) There is general agreement that administration of either thyroid extract, as was first demonstrated in the hen by Zawadowsky (1922), or thyroxine in the pigeon, brings on artificial molts, whatever the season may be. (2) On the other hand, surgical removal of the thyroid delays or even prevents the natural molt, providing that thyroidectomy occurs long before the normal onset of molt. The same occurs after chemical suppression of the thyroid gland. (3) A number of investigators have produced evidence of cytofunctional activation of the thyroid before the normal onset of molt. Using the goitrogen prevention method, Reineke and Turner (1945a) demonstrated a maximum in secretion of thyroxine during the molting period in the hen. On the other hand, whereas Tanabe et al. (1957) were unable to observe any increase in thyroid ^{131}I uptake in molting hens, Astier et al. (1970) have shown that, in the duck, a seasonal peak in thyroid ^{131}I uptake and in the ^{131}I conversion ratio (protein bound ^{131}I) occur within the month prior to the onset of the annual molt.

The thyroid gland seems thus definitely involved in molting by its stimulating effect on the development of young feathers, albeit that other endocrine glands, such as the gonads, seem to contribute to the complex endocrine mechanism of feather shedding and replacement, possibly by exerting a protecting effect on the old feather. Whether one or another of these glands plays the major role in the onset of molt or eventually some balance between hormones (e.g., thyroxine and sex steroids, or TSH and gonadotropins), needs further clarification. However, it can be stated here, that experimentally forced molt that always occurs in ducks some weeks after pituitary autograft, i.e., at a time when the circulating androgens fall to near zero values while thyroid hormones remain much less affected, can be consistently prevented either by moderate (0.4 mg/day) testosterone treatment, or by feeding propylthiouracil.

Reproduction

For several decades a number of converging investigations have demonstrated pronounced beneficial effects of thyroid hormones on gonadal function in domestic birds. However, Woitkewitsch's (1940) observations on inverse thyroid-gonad relationships in the European Starling, which were followed by very intensive investigations on several species of subtropical Indian finches by Thapliyal's group (see below), have led us to distinguish at least two main categories of birds in regard to their thyroid-gonad relationships: (1) birds that require thyroid hormones for a normal gonadal development and (2) birds that may achieve gonadal development without the thyroid glands.

The domestic species obviously belong to the first group. In male domestic birds, injections of moderate doses of thyroid hormones promote growth of the testes in young chickens, in adult domestic fowls, and in drakes, whereas thyroidectomy inhibits testicular development in young cockerels and ducklings blocks the phototesticular response in adult ducks and causes gonadal regression in cockerels. Although their effects are less constant, goitrogens may also lead to depression of spermatogenesis, comb size, fertility in cocks, and to inhibition of the phototesticular response in adult ducks.

Several wild species exhibit similar thyroid-gonad relationships. Thyroid hormones induce testicular growth "off-season" in House Sparrows (*Passer domesticus*), while in the Baya Weaver (*Ploceus philippinus*) the gonadal cycle remains suppressed indefinitely after thyroidectomy.

In female domestic fowl, thyroidectomy also induces delayed gonadal maturation and decreases egg production, and thiouracil treatment reduces ovarian weight in ducks and egg production and fertility in hens. However, thyroid extracts exert marked inhibition on the ovary of pigeons that are photostimulated or treated with pregnant mare serum and on light-stimulated Japanese White-eyes (*Zosterops japonica*), but the inhibition of ovarian growth may result from a depression of the concentration of yolk precursors for the ovary.

The assumption that in domestic species the thyroid exerts exclusively a promoting role on gonadal development was recently challenged when Lehman (1970) claimed that administration of a very low dosage of thyroxine to 2-week-old cockerels had no effect on testis weight and ^{32}P uptake, but lowered significantly the pituitary gonadotropic potency. On the other hand, administration of thyroxine, either during the vernal progressive phase or during autumn in ducks that were stimulated by artificial long days, had no marked effect on testis weight, but caused a severe decrease in plasma testosterone, associated with an augmented metabolic clearance rate of the hormone. Since similar alterations of testosterone production and metabolism occur normally in early summer, when the thyroid gland undergoes an annual peak of activity in ducks, the seasonal activation of the thyroid could appear at least as part of the complex mechanism whereby gonadal regression comes about in this domestic but seasonally cycling species.

The European Starling and several species of subtropical Indian finches belong to the second type of thyroid-gonad relationships, in which gonadal growth takes place even in the absence of thyroid hormones. Moreover, in such birds thyroid hormones play an essentially inhibiting role on gonadal (gonadotropic?) function. Regarding the effects of thyroidectomy, these birds fall into two subgroups. In the male Chestnut Mannikin (*Lonchura malacca*) and the Red Avadavat (*Amandava amandava*) removal of the thyroid gland before the onset of gonadal development leads to a marked extension (over 5 months) of the active phase of the gonad, resulting in a delayed and shortened regression phase. In the second subgroup of birds, including the females of the Baya Weaver, the Chestnut Mannikin, and both male and female Nutmeg Mannikin, the gonadal cycle of thyroidectomized birds is fixed at a maximum level for an indefinite period (more than 3 years), while regression can be precipitated any time of the year by thyroxine administration. Since thyroidectomized starlings were not followed over year-long periods, they cannot be aligned with either of the subgroups.

Several patterns of interactions of the thyroid gland on the reproductive cycle thus seem to exist, and a deeper insight into these relationships obviously will depend on more accurate measurements of blood concentration of thyroid hormones throughout the reproductive cycle of the different species.

Migratory behaviour

The problem of the relation between thyroid activity and migration has been raised repeatedly; Hacker (1926), Kuchler (1935), and Merkel (1938) showed that migrating birds have very active thyroid glands, while resting individuals have inactive glands. Merkel (1938) also was the first to simulate migratory behaviour (*Zugunruhe*) with injections of thyroxine or TSH.

Regulation of Thyroid Function

Hypothalamic-hypohyseal control

Pituitary control. Adenohypophysectomy in birds elicits marked atrophy of the thyroid gland. Several functional investigations have confirmed the considerable impairment of the thyroid function.

1. Hypophysectomized chickens on an iodine-supplemented diet (+2 μg/gm), 6 weeks postoperative, had approximately one-third as much thyroid ^{127}I as corresponding intact ones, but the same amount as either intact or hypophysectomized chickens on a low-iodine diet (0.14 μg/gm). On the other hand, in chickens on the low-iodine diet, hypophysectomy produced a 65% decrease of thyroid [^{127}I] thyroxine, with a twofold increase of [^{127}I] diiodotyrosine. No such effects were observed in iodine-supplemented hypophysectomized chickens, the controls of which having lowered thyroid iodine content, as was stated above. In hypophysectomized ducks, the thyroid content in ^{127}I was also found decreased, but the effect was less striking than in chickens.
2. Hypophysectomy markedly impairs uptake of ^{131}I by the thyroid gland. In growing chickens, this parameter of thyroid function, was reduced 6-8 weeks postoperative to 15% of controls, 19 hours after injection of ^{131}I, and to 20-25% of controls, irrespective of the amount of iodine in the diet, 24 hours after the tracer injection. In 2-3-week hypophysectomized ducks, a number of investigations indicated an ^{131}I uptake of 3% of controls, at 1 hour, which did not rise above 10% of controls 72 hours after injection. In Japanese Quail, 2-3 weeks after the operation, the corresponding values were 7% of controls at 12 hours and 22% at 72 hours. Under similar conditions, pigeons 12-25 weeks postoperative had an ^{131}I uptake of 15% of controls at 24 hours, and of 66% at 72 hours after the tracer injection. When measured *in vivo* in hypophysectomized pigeons, the thyroid radioactivity curve was also markedly depressed, the maximum of radioactivity being only one-third of the control maximum, and this peak was attained 19

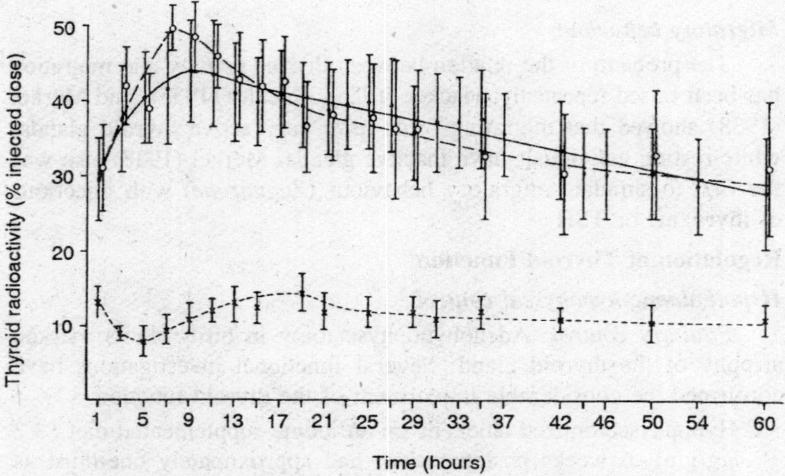

Fig. 3.6. In vivo measured radioactivity of thyroid glands in ^{131}I-injected pigeons, after hypophysectomy (×---×) or pituitary autografting (o--o) and in control (●—●).

hours after injection of ^{131}I in hypophysectomized birds compared with 9 hour controls. It is interesting to note that the depression of the thyroid ^{131}I uptake after hypophysectomy is observable in the chick embryo.

3. Radiochromatographic studies, on the other hand, have shown that in the hypophysectomized chicken, regardless of the level of iodine in the diet, 24 hours after injection of radioiodine, the extent of labeling of the thyroid pool of thyroxine amounted to one-tenth that of intact birds, while the percentage of diiodotyrosine was increased. Under similar conditions, the percentage of ^{131}I in thyroxine was 15-20% in ducks, 10% in pigeons, but 64% in the Japanese Quail.
4. At the peripheral level, the plasma protein-bound ^{127}I was lowered to 21% of controls in hypophysectomized chickens on low iodine intake, and to 35% of controls in iodine-supplemented birds. Furthermore the ^{131}I conversion ratio curve (protein-bound ^{131}I) also showed severe depression. In hypophysectomized ducks, the protein-bound ^{131}I was 20% of controls 24 hours after injections of ^{131}I, reaching 50% of controls at 72 hours. For hypophysectomized pigeons and Japanese Quail, the corresponding values were 15% of controls at 24 hours, and, respectively, 55% and 85% of controls at 72 hours.

5. Finally, measurements of the biological half-life of [^{125}I] L-thyroxine in hypophysectomized ducks have shown a marked depression of peripheral metabolism of the hormone; the half-life was 5.05 hours compared with 2.10 hours in controls. The delayed appearance of protein-bound ^{131}I within the plasma may thus reflect the reduced output of thyroid hormone, as well as the reduction of peripheral metabolism.

All these findings, which point to the preeminently controlling role of the pituitary in thyroid function in birds, are reminiscent of similar conclusions for mammals. No significant difference can be detected between these classes of vertebrates in this respect.

Hypothalamic control. Although the hypothalamus has an important role in the control of the thyrotropic function of the anterior pituitary, the intimate vascular relationship between them appears less important than in mammals. Thus, whereas ectopic pituitary autografting improves rather moderately thyroid function of hypophysectomized mammals, recovery is much more extensive in birds. (1) Although thyroid weights may still be reduced in autografted pigeons and Japanese Quail, atrophic glands are found in neither ducks or chickens with sectioned hypophyseal portal veins, pituitary autografted ducks, nor chickens. (2) The thyroid glands of ducks with autografts or severed portal vessels exhibit iodine accumulation, as indicated by high ^{127}I content. And despite the fact that thyroid-^{131}I uptake is scarcely affected in the autografted chicken, in ducks with autografts or severed hypophyseal portal veins, and in pigeons with pituitary autografts, a more detailed analysis of thyroid function in these three species has revealed a decreased rate of iodine metabolism. (3) The relative thyroid content in thyroxine-^{131}I is always significantly depressed 24 hours after ^{131}I injection. Subnormal values may also be observed after 48 hours. (4) Correlatively, the ^{131}I conversion ratio (protein-bound ^{131}I) was generally depressed during the first 24 hours after the tracer injection. (5) The half-life of [^{125}I] thyroxine in ducks with pituitary autografts also showed an appreciable restoration of peripheral metabolic rate of the hormones; the half-life was 3.13 hours, compared with 2.10 hours in controls and 5.05 hours in hypophysectomized birds.

In summary, the basal pituitary thyrotropic function of birds has a fair degree of autonomy with respect to its hypothalamic control. However, the absence of the normal hypothalamo-hypophyseal connection does lead to reduction of metabolism in the thyroid gland, even when the animal is at the basal state. The role of the hypothalamo-pituitary axis in the regulation of thyrotropic function by environmental factor will be discussed anywhere. But it may be mentioned here that ducks with disconnected hypophyseal veins fail to exhibit goitrogenic reaction to propylthiouracil treatment.

Rhythmicity of thyroid function

A number of morphological investigations on the avian thyroid glands have led to the assumption of the occurrence of seasonal variations of thyroid function. Several observers have claimed that histological stimulation of the thyroid gland coincides with the onset of the cold season, e.g., Riddle et al. (1932) in the pigeon; Burger (1938) in the European Starling; Hohn (1949, 1950) in the Mallard; Davis and Davis (1954) in the House Sparrow; Oakeson and Lilley (1960) in *Zonotrichia leucophrys nuttalli*; Wilson and Farner (1960), Farner (1965) in *Zonotrichia leucophrys gambelii*; Legendre and Rakotondrainy (1963) in the Red Fody (*Foudia madagascariensis*).

On the other hand, Bigalke (1956) and Fromme-Bouman (1962) have shown, in the European Blackbird, a morphological activation in thyroid in May, whereas the Nutmeg Mannikin exhibits an annual thyroid cycle with morphological indications of lower activity in August-September. Now a systematic month-to-month study of the annual cycle of thyroid-^{131}I uptake and of plasma protein-bound ^{131}I 24 hours after injection of radioiodine has been carried out in male ducks reared under field conditions within the mild climatic conditions of the Mediterranean region. An annual peak of both parameters occurred in June, together with an increased relative thyroid content of [^{131}I] thyroxine 24 hours after injection of ^{131}I and with a decreased thyroid ^{127}I content, indicating a marked thyroid activation at that period. In a similar study on the Nutmeg Mannikin, Chandola (1972) noted a depressed thyroid ^{131}I uptake in August-September (breeding season) as compared with other months.

These observations indicate that the avian thyroid gland has seasonal functional fluctuations. However, they do not provide decisive explanations of the possible factors involved herein, such as the ambient temperature, the photoperiod, and the sexual cycle, all of which will be discussed in the following sections.

Finally, it must be stated that up to now no diurnal or circadian rhythms in thyroid function have been identified in birds. The only statement on short periodic functions is that of a 12 hour rhythmicity in thyroid ^{131}I uptake recently demonstrated in the duck.

Gonadothyroid relationships

The annual cycle in thyroid activity has been related to the reproductive cycle. There is morphological evidence of depression of thyroid function during the reproduction period in *Zonotrichia leucophrys nuttalli*, *Foudia madagascariensis*, *Lonchura punctulata*, and in most finches studied by Thapliyal (1969). In other species, such as the European Starling, the Mallard, and the European Blackbird, the thyroid gland goes through a stage of stimulated activity as the gonads pass into their seasonal regression. The same pattern has been demonstrated with radioisotopic techniques in the duck and in *Lonchura punctulata*. Such inverse phase relationships between the seasonal cycle of the thyroid gland and the photoperiodically induced sexual cycle could be relevant to gonadothyroid relationships and/or to photoperiodic regulation of thyroid function.

Although gonadothyroid relationships undoubtedly exist, at least in the male, the precise mechanism is still controversial. In the male, Kobayashi (1954a) observed stimulation of the thyroid parenchyma in testosterone-treated pigeons, and Stetson and Erickson (1971) found no effect on thyroid weight in castrated White-crowned Sparrows (*Zonotrichia leucophrys gambelii*) subjected to either long or short days. On the other hand, capons exhibit greater thyroid weight than intact chickens, but Oddel (1952) found no modification of the thyroxine secretion rate in castrated chickens. Chandola (1972) measured an increased thyroid ^{131}I uptake in *Lonchura punctulata* receiving testosterone, whereas castration had the opposite effects. In the duck, castration has been found to increase the circulating TSH level, together with the pituitary thyrotropic cells, to increase thyroid weight, thyroid-^{131}I uptake, and protein-bound ^{131}I, while testosterone treatment had the opposite effects on both thyroid radioiodine tests, in both castrated and intact birds. A similar depressive effect of testosterone was observed on pituitary thyrotropic cells of the Japanese Quail. However, at the peripheral level, testosterone accelerates the metabolism of thyroxine, decreasing its biological half-life in the duck.

The seasonal decrease in testosterone secretion could thus be at least partially responsible of the annual increment in thyroid activity in ducks. However, since, vice versa, hyperthyroidism also inhibits testosterone secretion (*vide supra*), it appears difficult, at the present time, to assess which of the endocrine alterations, hyperthyroidism or hypoandrogenism, may be considered as the primary inducer of this peculiar seasonal endocrine balance.

Less attention has been given to gonadothyroid relationships in females. Estrogens have been claimed to have no effect on thyroid weight in chickens but to depress the thyroid follicles in pigeons and to lower the thyroid ^{131}I uptake in *Lonchura punctulata*. Progesterone induces marked inhibition of thyroid-^{131}I uptake in hens and in Bengalese Finches (*Lonchura striata*).

Environmental factors

Light and day length

Despite the occurrence of definite and probably inhibitory repercussions of photoperiodically stimulated gonadal activity on the thyroid gland of most species light has been claimed to act somehow as a stimulator of thyroid function. This statement is mainly based on histological observations of thyroid glands of European Starlings and of ducks and of the hypophyseal thyrotropic cells of ducks and of Japanese Quail. Light has also been shown to enhance thyroxine-secretion rate in the chicken. However, radioisotopic investigations have led to more ubiquitous conclusions.

1. An experimental long-day regimen depressed markedly thyroid ^{131}I uptake in intact drakes and in male and female Japanese Quail, in which photostimulated sexual steroids could be interacting, as well as in castrated ducks.
2. The same investigators also observed a mild depression of the conversion ratio (protein-bound ^{131}I) in intact or castrated ducks and in male Japanese Quail. However, Follett and Riley (1967) concluded from an increased slope of the protein-bound ^{131}I within the first hours after the injection of tracer into female quail that the hormonal output from the thyroid was stimulated somewhat by the light regimen.
3. Long days exert a slight inhibition on peripheral thyroxine catabolism evaluated by the half-life of thyroxine in castrated ducks, and inhibit the above-mentioned decrease in thyroxine half-life due to exogenous testosterone in intact controls.

Although no decisive conclusions result from these observations, the radioiodine tests bring more arguments in favour of an inhibition of thyroid activity than of stimulation by light, at least in birds with photoregulated gonadal cycles. Of special interest are, therefore, the recent investigations with similar methods on Nutmeg Mannikins, which are not dependent on photoperiod for the regulation of the reproductive cycle. In this species long-day photoperiods, that did not lead to gonadal recrudescence, induced a significant stimulation of the thyroid ^{131}I uptake and of the plasma protein-bound ^{131}I, indicating a definite activation of thyroid metabolism.

In addition to the problem of the eventually stimulating or inhibiting action of the photoperiod on thyroid function and metabolism, another unanswered question concerns the pathway whereby light may act on the thyroid gland. It is tempting to admit *a priori* that a route similar to that of the photoneuroendocrine mechanism controlling the gonadal cycles of many avian species may also convey photostimuli to the thyroid gland. Alternatively some neural or metabolic factors linked to locomotor activity, which is definitely controlled by the light-dark cycle, could be involved in this mechanism. The maintenance of the depressive effect of long days on ^{131}I uptake and protein-bound ^{131}I in hypophyseal autografted quail could actually point to the latter hypothesis.

Temperature

Several previously cited investigations on seasonal fluctuations of thyroid function have reported an annual maximum of thyroid activity during the cold season. In less systematic studies, comparing winter to summer trials in domestic fowl, it has been stated that thyroid glands were larger in winter than in summer. The thyroid iodine content was equally depressed in summer, while the thyroxine secretion rate declined about 15% from March to May, the total reduction from winter to summer reaching 58%. Thyroid ^{131}I uptake in ducks also was found higher during the coldest months (January-February) than under the milder March temperatures. Similarly, Kobayashi et al. (1960) noted in the White-throated Sparrow a higher ^{131}I uptake associated with increased thyroid [^{131}I] thyroxine levels in February than in May. On the other hand, the biological half-life of [^{131}I] thyroxine, as measured from the decrease of the total plasma radioactivity, was found to be lower in winter in domestic fowls (7.6 $\pm$ 0.50 hours) than in summer birds (11.4+0.50 hours).

The stimulating effect of low environmental temperature on thyroid function thus demonstrated has been studied experimentally. Cockerels kept for 3 weeks at 7.5°C had higher acinar-cell levels and higher thyroxine secretion rates (goiter-prevention method) than others kept at 23.5° or 31.5°C. Similarly, pullets reared at 13°C had a greater thyroxine-secretion rate (radioiodine method) (3.47 μg/100 gm) than at 29.5°C (2.59 μg/100 gm), and ducks exposed to 7.2°C exhibited an enhanced rate of thyroxine secretion (thyroxine-degradation method) (2.05$\pm$0.17 μg/100 gm per day), compared with birds, at 25°C (1.57$\pm$

0.05 μg/100 gm per day). Indirect evaluations of TSH secretion rate also indicated higher values in lower as compared to higher environmental temperatures. Acute cold exposure (4.4°C for summer temperature-adapted chickens) caused increased rates of uptake and release of thyroid ^{131}I after at least 24 hours. Both parameters reached a maximum after 72 hours and returned to normal in a longer term (1 month) of exposure to cold. The time required for thyroid stimulation by cold exposure thus appears to be much longer in birds than in mammals.

The effect of acute cold exposure (24 hours from 25°C to 4°C) has also been investigated in adult ducks by measuring the equilibration rate of radioactive [^{131}I] iodoamino acids (monoiodotyrosine, diiodotyrosine, thyroxine) with the respective ^{127}I components in thyroid extracts after ^{131}I injection. Equilibrium was reached after 48 hours in the cold-exposed individuals, as well as in the controls. However, the slope of the specific radioactivity curve of the total iodine content of the thyroid gland, of mono- and diiodotyrosine and thyroxine was steeper in the cold-exposed animals, the corresponding maxima were higher, and the maxima were attained earlier (14 hours compared with 24 hours) in the cold-exposed ducks than in the controls. On the other hand, adult ducks exposed for 30 days to 5°C exhibited decreased thyroid [^{127}I] thyroxine (0.39 ± 0.02 μg per milligram thyroid compared with 0.59 ± 0.03), decreased protein-bound ^{127}I (1.71 ± 0.07 μg per 100 ml plasma compared with 1.90 ± 0.06), and decreased half-life of [^{125}I] thyroxine (2 hours 42 minutes compared with 3 hours 23 minutes), compared to the controls kept at 23°C. Similarly, McFarland et al. (1966) measured a decrease in half-life of the radiothyroxine (18.4 hours) in cooled (21°C) Japanese Quail, as compared to birds kept at warmer temperatures (32°C for 30.4 hours), while Hendrich and Turner (1967) claimed that reduction of a mean winter environmental temperature at 12.8°C or a warm summer temperature (32.3°C) to a constant 4.4°C had no significant effect on the half-life of [^{131}I] thyroxine in the domestic fowl.

The opposite effect of change in temperature, namely, from cold to warmer environments, has received less attention. However, the return of long-term cold-adapted chickens (2 years in 4.4°C) to a warm environment (12.8-23.9°C) resulted in a rapid reduction of thyroxine secretion rate (from 0.91 to 0.53 μg per 100 gm).

The effects of high environmental temperatures on thyroid function have also been studied in the domestic fowl. Chickens maintained at

31.5°C showed reduced thyroid weights, thyroid-^{131}I radioactivities, and thyroid-secretion rates, together with reduced oxygen consumption, than controls kept at 19°C. Similarly, cockerels grown for 3 weeks at 24°C had a thyroid-secretion rate (goiter-prevention method) of 3.65 μg/bird/day, compared with 1.60 μg/day at 35°C, and 0.7 μg/day at 40.5°C, the last animals being in hyperthermia.

Feeding

In contrast to mammals, in which emphasis has been placed on the depressive effects of malnutrition on thyroid function, little attention has been given to the thyroid repercussions of caloric malnutrition in birds. Underfeeding has been claimed to have no influence on relative thyroid weight in the chicken, although the metabolic rate is depressed. On the other hand, starvation for 5 days was found to inhibit completely thyroid release of ^{131}I in the domestic fowl, and food restriction inducing a 23% body weight loss also reduced the thyroid secretion rate to half normal values in the pigeon. However, in order to avoid interference with thyroid function by caloric malnutrition and iodide deprivation due to restricted food intake, more systematic investigations were performed in ducks previously adapted to an iodine-deficient diet. After 17 days of severe malnutrition (food intake reduced to 50% or 25% of controls or starvation) there was no evidence of marked impairment of ^{131}I uptake by the thyroid gland and of ^{131}I distribution among the intrathyroid amino acids. However more accurate methods of investigation such as the measurement of the "specific activity" [ratio of the labeled (^{131}I) to the stable (^{127}I) compounds] of the intrathyroid amino acids have yielded evidence of a depressed metabolism of the thyroid gland in undernourished ducks. Starvation delayed the peak values of the specific activity of the iodotyrosines from 6 hours to 24 hours, and that of the iodothyronines from 8 hours to 48 hours. In the plasma the ^{131}I conversion ratio (protein-bound ^{131}I) was markedly depressed in proportion to the extent of inanition. There was also a decrease in the plasma [^{131}I]-thyroxine level 18 hours after ^{131}I injection.

Endocrine Pancreas

Morphology

The avian pancreas is an elongated gland that lies primarily between the two limbs of the duodenal loop. It consists of two main lobes, the dorsal and the ventral lobes, which may be independent or fused, and a third, smaller splenic lobe, that extends, from the dorsal lobe toward the spleen. There are usually three secretory ducts that lead to the ascending limb of the duodenal loop.

The pancreas originates from three primordial evaginations of the gut, one dorsal rudiment, and two ventral anlagen, the dorsal evagination usually appearing first. The distal portion of the three rudiments proliferate masses of glandular tissue that later fuse to form the definitive pancreas. Histologically, the primordium of the pancreas first appears as a compact mass of tissue that differentiates later into cellular cords that gradually acquire lumina. The acinic cells, as well as the endocrine islets, differentiate from the trabeculae, or from the walls of the tubules throughout the entire organ, beginning at an early stage (8 days in the chick).

Light microscopy

In adults, the insular endocrine tissue generally is more abundant in the dorsal than in the ventral lobe; moreover, it is the main component of the splenic lobe. As in most vertebrates, the avian islet of Langerhans appears in two dimensions as a roughly circular aggregate of cells with a diameter ranging from 50 to 200 μ. The endocrine cells are accompanied by a thin layer of connective tissue that extends into the islet, dividing it into lobules. The majority of islets cells contain varied numbers of small granules that stain (e.g., with aldehyde-fuchsin and counterstain), and according to the cells, either blue (β, or B cells), or red to pink (α, α_2, or A cells), or sometimes orange to gray (α_1 or D cells). A few cells contain cytoplasm free of either granules (C cells). The arrangement of the two main cell types exhibits in birds a very peculiar pattern, since two types can be distinguished: the dark islands, containing mainly α cells, and the light islands, containing predominantly β cells.

Electron microscopy

Bjorkmann and Hellman (1964), Sato et al. (1966), and others have shown that α and β cells are clearly identifiable in all species. They are already differentiated by the day of hatching. In general, the β cells appear spherical to ellipsoidal and contain fewer aggregations of secretory granules than the α cells. An inverse relationship generally exists between number of secretory granules and number of other cytoplasmic organelles (unattached ribosomes, mitochondria, Golgi apparatus, and endoplasmic reticulum). In turn, the α-cell cytoplasm is characterized by a predominance of secretory granules, and a smaller proportion of other organelles. In both cell types, the granules are formed in the Golgi zone. β Granules are less electron dense, less homogeneous, and slightly larger than the a granules in all species studied. In the domestic fowl, most granules are present as bar-shaped crystalloids, whereas those of the turkey appear as numerous overlapping rods. Most β granules lie within a membrane-enclosed space. As in other vertebrates, the avian α granules appear spherical. The lining membrane also is more closely apposed to the granule surface than in β granules. Many β granules and a few α granules are surrounded by fine filaments. Dense multi-vacuolated cytoplasmic bodies 0.5-0.8 μ in diameter are present in both β and α cells.

Hormones

As in other vertebrates, the avian pancreas produces at least two hormones—insulin and glucagon.

Insulin

Emphasis has been placed on the low yield of insulin extractable from the avian pancreas in comparison with mammalian glands (1-2 mg insulin per 100 gm of pancreas in chicken, as compared with 10-15 mg/100 gm in the steer). Nevertheless, crystalline insulin has been recently isolated from chick, and turkey pancreas. Its amino acid sequence is close to that of pig and dog insulin, except for the amino acids in position 8, 9, and 10 in chain A, and 1, 2, and 27 in chain B. This difference in structure between chick and mammalian insulin may conceivably account, at least partially, for the much higher effectiveness of the avian hormone in promoting hypoglycemia in intact chickens, when compared with equivalent amounts of mammalian hormone. In fact, 0.02 units/kg produced a degree of hypoglycemia equal to that observed in chickens injected with a beef-insulin dose that is tenfold greater, and vice versa, mammalian (beef) insulin causes a stronger hypoglycemia in intact rats than does chicken insulin. On the other hand, *in vitro* assay systems appear to respond equally well to all

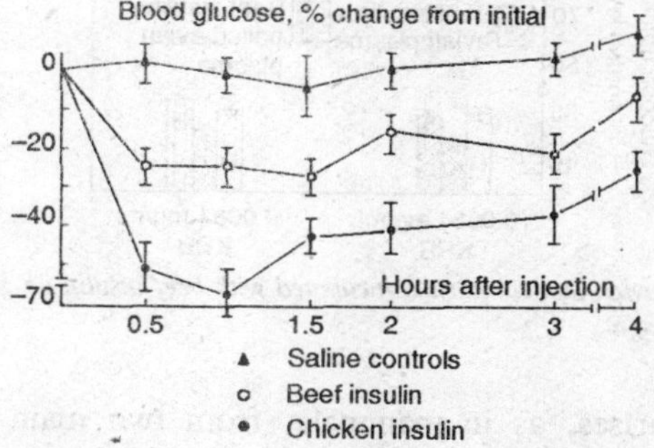

Fig. 3.7. Effect of intravenous injection of chicken and beef insulin (0.1 units) on the blood-glucose levels of the chick.

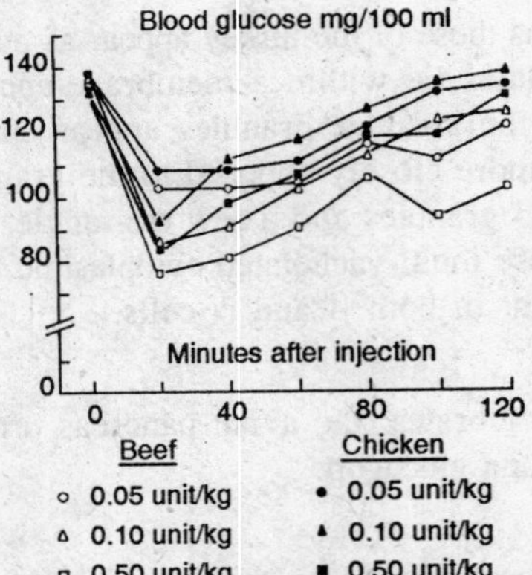

Fig. 3.8. Effects of intravenous injection of beef and chicken insulin on the blood-glucose levels of normal nonfasted rats.

insulins. More recently, some light has been thrown on the possible cause of such interspecific resistance toward heterologous insulins. In this experiment, preincubation of beef insulin with chicken plasma led to severe inhibition of glucose uptake by the isolated rat diaphragm, while preincubation of chicken insulin with rat plasma similarly reduced glucose uptake potential *in vitro*. These findings suggest that, in addition to differences in the amino acid sequence of insulins, and hence interspecific receptor-site differences, the occurrence of inhibitory plasma factor(s) may play a role in determining species specificity in the function of insulin. The normal plasma-insulin level (I.R.I., immunoreactive insulin) is about 18 μ unit/ml in the duck.

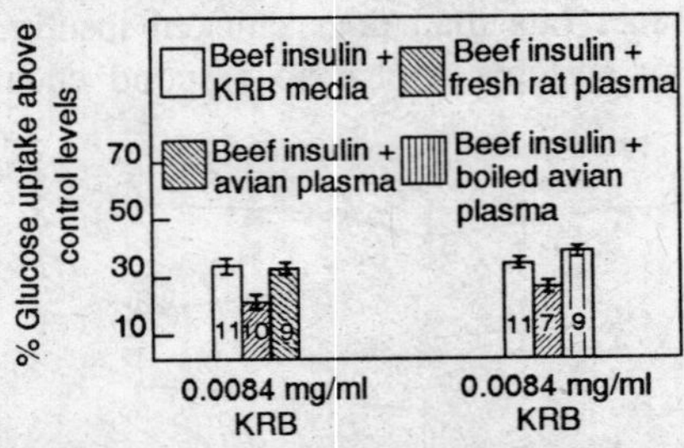

Fig. 3.9. Effects of avian and rat plasma incubated with beef insulin on glucose uptake by rat diaphragm.

Glucagon

Glucagon arises, as in mammals, from two main sources, the pancreas and the gut. The hormone has been extracted from the pancreas of the pigeon, the domestic duck, two carnivorous species,

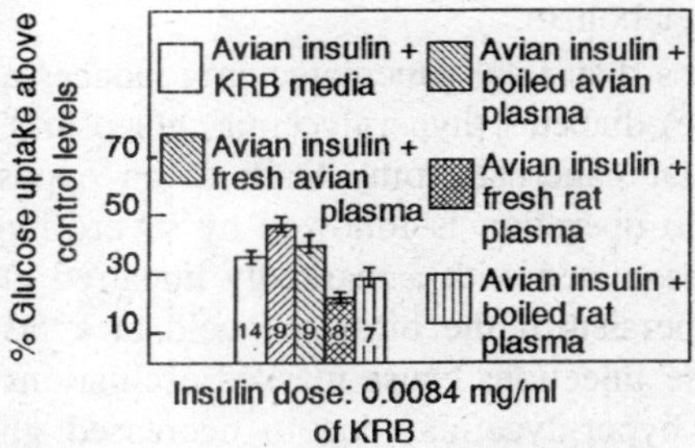

Fig. 3.10. Effects of avian insulin incubated with avian or rat plasma on glucose uptake by rat diaphragm.

the Barn Owl (*Tyto alba*), and the Common Kestrel (*Falco tinnunculus*) by Vuylsteke and de Duve (1953); almost ten times as much glucagon was found as in the pancreas of mammals. From recent biochemical studies on duck glucagon it appears that two forms of glucagon are found in the pancreas, one with a molecular weight (3000) equivalent to crystalline mammalian glucagon, and the other with approximately double that molecular weight, which seems similar to that secreted in very small amounts by the dog pancreas. Intestinal (jejunum or ileum) extracts contain large amounts of the large-molecule glucagon, but none of the smaller type, which contributes to the total glucagon level in the blood (1 ng per ml plasma in the fasting duck, e.g., more than four times the level found in fasting humans). In serum, the only large glucagon (6000) has been observed in either normal, totally pancreatectomized, or eviscerated ducks. Despite the fact that glucagon from both origins share some biological properties, the role of gut glucagon seems to be of restricted importance, since its plasma concentration is rather low.

Role of Pancreatic Hormones

Carbohydrate metabolism

For years, the endocrine pancreas has been considered to be of limited importance in the regulation of carbohydrate metabolism in birds, since a number of investigators have claimed that pancreatectomy in the chicken, pigeon, duck, and goose induced either no diabetes at all, or at most, a mild diabetes lasting 1 week without any further alteration of carbohydrate metabolism. This classic assumption has been questioned by the more recent and extensive investigations of Mialhe (1958, 1969, 1971) in the duck, Sitbon (1967), and Karrman and Mialhe (1968) in the goose, and Mikami and Kazuyuki (1962) in the chicken. The data from these experiments leave no doubt concerning the preeminent role of the two pancreatic hormones in avian carbohydrate metabolism.

Whereas less than total pancreatectomy induces a transient (duck) or severe (goose) diabetes (hyperglycemia, glucosuria, polydipsia, and polyphagia), total pancreatectomy leads to an opposite situation. In both species, the operation is followed by severe hypoglycemia (and convulsions), associated with a markedly impaired glucose tolerance. Hypoglycemia persists if the birds are held in a fasting state. Food intake or glucose injections cause marked oscillations of blood sugar, from hypo- to hyperglycemia, due to decreased glucose tolerance. Survival may also be prolonged by administration of hyperglycemic hormones such as glucagon, adrenaline, or ACTH. In the fasting pancreatectomized duck, the blood glucose level can be maintained in normal range by repeated injections of glucagon and insulin in a w/w ratio of 1.5:3. Glucose tolerance is still impaired, but can be restored to normal by increasing the insulin dose.

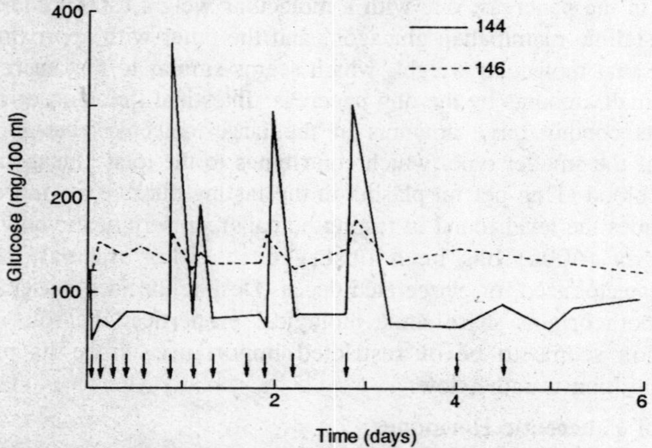

Fig. 3.11. Hypoglycemia and impaired glucose tolerance after total pancreatectomy in the duck.

Mikami and Kazuyuki (1962) have made observations on the domestic fowl that similarly emphasized the role of glucagon. They showed that extirpation of the two lobes of the pancreas that contain the a cells induced severe hypoglycemia, which could be restored by glucagon injections, whereas partial pancreatectomy involving other lobes of the gland elicited a mild and transient hyperglycemia.

The hypoglycemic role of insulin and the hyperglycemic effect of glucagon have now been amply verified not only in domestic birds, such as domestic fowl and ducks and geese, but also in finches. The glycogenic effect of insulin together with the glycogenolytic effect of glucagon have been demonstrated in pancreatectomized ducks and in intact fowls. It can, therefore, be concluded that:

1. In fasting pancreatectomized birds, hypoglycemia is due to a lack of glucagon, and decreased glucose tolerance is due to a lack of insulin. The discrepancy between recent and earlier findings on pancreatectomized birds presumably is due to incomplete extirpation of the gland by the earlier investigators. However, further investigation will be needed to throw light on the behaviour of pancreatectomized carnivorous species, such as owls, which have been claimed to become definitively diabetic after pancreatectomy.
2. In the normal bird, as in mammals, glucagon seems to play an important role in maintaining a normal blood sugar level in the fasting state, whereas insulin prevents hyperglycemia after feeding.

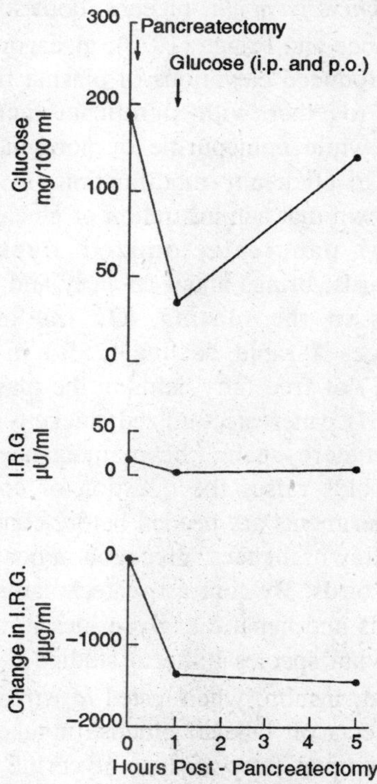

Fig. 3.12. Total pancreatectomy in duck causing a large decrease in plasma glucose, immunoreactive glucagon (I.R.G.) and immunoreactive insulin (I.R.I.).

3. The ratio of the plasma concentrations of glucagon and insulin (G/I) seems to be of major importance for the regulation of the blood-sugar level.

On the other hand, *in vivo* and *in vitro* studies have shown that the glycogen body, which is a small mass lying dorsal to the sacral spinal cord within the vertebral column and contains 60-80% glycogen of unknown role, fails to react to avian as well as to mammalian insulin.

Lipid metabolism

Recent investigators have emphasized the prominent lipolytic action of glucagon in birds. Glucagon has a marked stimulatory effect on the concentration of free fatty acids in the plasma *in vivo* in the domestic fowl, the duck and the goose and the turkey, as well as on the release of free fatty acids *in vitro* from adipose tissue in *Passer domesticus* and *Zonotrichia leucophrys gambelii*, pigeons, domestic fowl, and ducks.

According to Grande and Prigge (1970), glucagon infusion (2 hours) in geese and ducks produced elevations of plasma free fatty acids and plasma triglycerides together with significant increase of the liver triglyceride content, while epinephrine or norepinephrine infused at the same rate failed to elicit any modification. In ducks, Desbals et al. (1970a,b) have shown that administration of glucagon to either intact fasting, to totally pancreatectomized ducks, or even to hypophysectomized birds, brings about an early and important increase of free fatty acids in the plasma. On the other hand, total pancreatectomy induces a rapid decline (25%) in lipemia, together with a decrease (50%) of free fatty acids in the plasma. According to Lepkovsky et al. (1967) pancreatectomized chickens would behave in a different manner, but here again, the maintenance of normal blood levels of free fatty acids raises the question of completeness of the operation; further experiments are needed before conclusions on species differences may be drawn. In fact, glucagon appears to be the major adipokinetic factor in birds. By contrast, catecholamines, which elevate plasma free fatty acids and hepatic triglycerides in mammals, have no such effects in the avian species hitherto studied.

On the other hand, insulin, when tested *in vitro* in passerine birds, has no detectable effects on labeled glucose uptake by abdominal or furcular fat pads, nor upon fatty acid or glyceride glycerol synthesis from labeled glucose by the adipose tissue. Similarly, fatty-acid synthesis in adipose tissue of the pigeon studied *in vitro* and *in vivo* was not appreciably influenced by insulin. Moreover, insulin was found to have no antagonistic (antilipolytic) effect on glucagon, either *in vitro* or *in vivo* in the duck and in the chick, nor did it counteract epinephrine-induced lipolysis in pigeon adipose tissue. However, in isolated chick adipose tissue, insulin stimulates the incorporation of glucose-U-^{14}C into glyceride glycerol.

In vivo insulin injections into intact fasting ducks, domestic fowl, and ducks reduce lipemia but induce a delayed (30 minute) increase in free fatty acids in plasma. This might be attributed to an indirect release of glucagon, as the latter effect does not occur in pancreatectomized ducks.

Regulation of Pancreatic Hormone Secretion

Glucose

As in mammals, blood-sugar level serves in birds as a major regulator of secretion of pancreatic hormones. Acute or chronic glucose

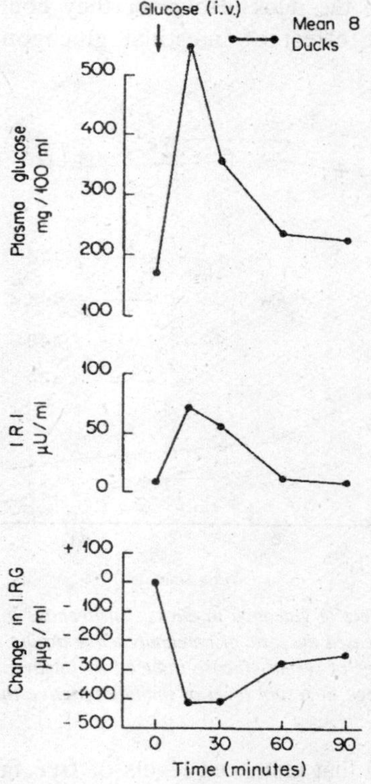

Fig. 3.13. Effect of intravenous glucose (1.75 gm/kg body weight) in ducks.

injections, or intrajejunal infusions of glucose, induce a sharp increase in immunoreactive insulin in the goose, in the duck and in the domestic fowl. In correlation, the concentration of immunoreactive glucagon in the plasma decreases following intravenous injection or intrajejunal infusion of glucose. The feedback mechanisms glucose-glucagon and glucose-insulin have been shown to apply to physiological variations of blood-glucose level.

Glucagon-insulin interactions

Measurements of immunoreactive insulin after injections of glucagon in ducks have provided direct evidence for a glucagon-induced insulin secretion. This insulinogenic effect of glucagon is even higher than that of glucose infusions, inducing similar increases in blood glucose. On the other hand, Samols et al. (1969b) have demonstrated a significant increase in concentration of immunoreactive glucagon after intravenous injection of insulin in the duck. However they could not observe a specific insulinogenic effect of intestinal glucagon in response to alimentary glucose.

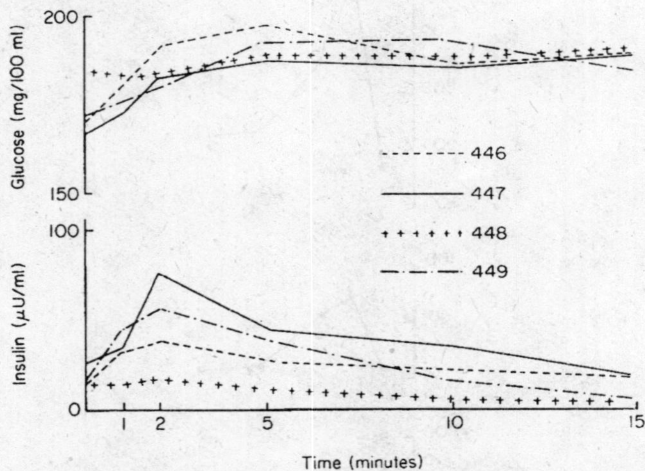

Fig. 3.14. Insulinogenic effect of glucagon in ducks. Intravenous injection of 1.25 μg/kg of glucagon increased the level of immunoreactive insulin in 3 out of 4 animals. With similar rises of serum glucose induced by intravenous glucose infusion, there is no change in serum level of immunoreactive insulin.

Free fatty acids

It has been stated that enhanced levels of free fatty acids may be a potent stimulator of plasma glucagon in the duck, a fact that enlarges the concept of metabolic feedback regulation of glucagon secretion.

Permissive role of other endocrines

As in mammals, pituitary or pituitary-controlled hormones have been shown in the duck to interfere with regulation by pancreatic hormones.

1. Hypophysectomy improves the glucose tolerance of totally pancreatectomized birds, while it enhances the hypoglycemic response to total pancreatectomy.
2. Hypophysectomized ducks are hypersensitive to insulin, i.e., they show a supernormal glucose tolerance with an insulin-induced hypoglycemia following the hyperglycemic peak. This hypersensitivity can be corrected by administration of either growth hormone or corticosterone.
3. Hypophysectomy markedly impairs the hyperglycemic effect of glucagon. The normal response is restored if the hypophysectomized animals are treated with growth hormone or corticosterone.
4. The anterior pituitary, or at least growth hormone, and corticosterone, are also necessary for a normal insulinogenic effect of glucagon.

Gastrointestinal hormones

In mammals, this peculiar group of hormones has a marked stimulatory effect on glucagon and insulin secretion. Pertinent data also are available for the duck, in which pancreozymin injections induce a definite and prolonged increase in glucagon secretion, whereas insulin secretion shows only a very transient stimulation.

Drugs

The administration of hypoglycemic "antidiabetic" sulphonylureas (e.g., tolbutamide) to ducks produces (1) a rapid decrease in plasma glucagon level and (2) a sharp increase in plasma insulin. As either effect may occur before any hypoglycemic response, and as in some instances, the suppression of pancreatic glucagon is associated with only negligible change in insulin levels, the drugs seem to have at least partly direct and independent effects on both major islet components. On the other hand, diazoxide, a potent hyperglycemic sulfamide in mammals, also induces hyperglycemia associated with a transient lowering of the plasma insulin in the domestic fowl.

Parathyroid Glands

Morphology

The parathyroids arise from the dorsal wings of the third and fourth branchial pouches that thicken very early into a solid mass of cells, the primordium of a parathyroid gland. Two glands are thus found on each side close to the posterior pole of the thyroid. The number and location of the four parathyroids remain unaltered in most species, although partial fusion of two homolateral glands may occur in some of them (domestic fowl). In addition to this general pattern, chickens often possess functional accessory parathyroid tissue, which is embedded within the ultimobranchial glands. This peculiarity may account for the occasional failure of parathyroidectomy to induce tetany as reported by several authors for this species.

Light microscope studies have revealed that the main difference between mammalian and avian parathyroid glands lies in the absence, in the latter, of oxyphilic cells between the chief cells. The chief cells are arranged in elongated or branching cords of polygonal to columnar cells, separated by a thin stroma of connective tissue, with numerous sinusoid capillaries. When active, the nucleus of the chief cells may be hyperchromatic and the cytoplasm very granular, and hence rather dark with few vacuoles. When inactive, the nucleus and cytoplasm may stain very lightly, and the poorly granulated cytoplasm contains numerous vacuoles. It is interesting to note that generally chief cells belonging to the same area of the gland exhibit the same functional characteristics, while different levels of activity may be observed within distant areas of the gland. Hyperplasia and hypertrophy of the parathyroids occur if the bird is deprived of dietary calcium or vitamin D, or of ultraviolet light. Pronounced hypertrophy of the gland also parallels maximum reproductive activity. On the other hand, fasting, aging, and hypophysectomy cause structural regression.

Electron microscopic study of the parathyroid gland of laying hens, i.e., of a presumably actively secreting gland, has recently been performed. The ultrastructural organization of the parathyroid cells corresponding to the mammalian chief cells appear to be, in principle, very similar to that of other vertebrates hitherto studied. The parenchymal cells show a tortuous plasma membrane, and the highly organized cytoplasm contains well-developed organelles and rod-shaped mitochondria, and vesicles and cisternae of the smooth endoplasmic reticulum are dispersed throughout the cytoplasm. Aggregations of free ribosomes and a well-developed rough endoplasmic reticulum, which consists of parallel cisternae with attached ribosomes, are also present. The Golgi apparatus is prominent and consists of dilated concentric cisternae and vesicles of variable size. Numerous small prosecretory membrane-bound granules 500 Å in diameter are observed in the Golgi area as well as in the cytoplasm outside the Golgi complex. A few large granules, 1000-4000 Å in diameter, which correspond to the mature secretory granules described in other vertebrates, are also observed in the cytoplasm, but no fusion of prosecretory granules to form larger droplets was seen. Finally, there are also in the cytoplasm coated vesicles that sometimes appear to be fused with the plasma membrane. The hormone is believed to be synthesized in the granular ergastoplasm and transferred to the Golgi apparatus, where it is packed into small prosecretory granules, and finally discharged into the intercellular space. The formation of the coalesced larger storage granules does not seem to be necessary.

Hormone

The parathyroid hormone, or parathormone, appears to be a straight-chain polypeptide. In mammals, parathormone has been claimed to contain 73 amino-acids residues, with a molecular weight of approximatively 8500, and a biological activity of 2500 U.S.P. units/mg, but little is known of the precise structure of the avian hormone.

Role

As in mammals, the vital importance of the parathyroid in birds is evidenced by the fact that the total removal of parathyroid tissue leads to death in about 24 hours. Clinically, death is due to tetany, and biochemically, the main syndrome lies in hypocalcemia, with blood calcium concentrations reaching 5 mg% in about 20 hours instead of 10 mg% in the normal bird, and is associated with increased phosphoremia.

On the other hand, despite possible species differences in the biochemical structure of avian and mammalian parathormone, injections of mammalian parathormone have been shown to induce hypercalcemia in the duck and in the fowl.

Calcium homeostasis is the major function of the parathyroid gland. Calcium ions are involved in many functions, such as regulation of neuromuscular irritability, muscular contraction, modifying the permeability of capillary membranes to water, clotting of blood, calcification of bone and of eggshells, and as a cofactor of various enzymes. The parathyroid glands control the plasma concentration of calcium ions, which in turn is in equilibrium with the protein-bound blood calcium. Plasma calcium is adjusted principally by extraction of the calcium from the skeleton, control of the renal excretion of calcium phosphate, and increase of the rate of absorption of calcium by the gut. Emphasis has been placed on the synergistic action of parathormone and vitamin D, which also increases the rate of absorption of calcium from the gut, together with the rate of resorption of bone calcium. In fact, bone is the primary target organ of parathyroid hormone. Parathyroid hormone-induced hypercalcemia may depend on the resorption of metaphyseal and endosteal bone tissue, the skeleton acting as the most extensive calcium reservoir of the organism. Moreover, the parathormone has been found to induce differentiation of a significant number of active osteoclasts, which contribute to resorption of bone tissue.

Calcium homeostasis and parathyroid function exhibit marked activation in female birds during the reproduction cycle. Eggshell production requires an adequate calcium supply to the shell glands of the uterus. How the movement from the plasma through the shell gland is regulated is not yet clear, although there is some evidence that the shell gland acts as a calcium pump involving a protein-binding transfer mechanism similar to that of the gut. By this mechanism, calcium is removed from the blood during the main period of eggshell formation (16 hours in the fowl) at a mean rate of approximately 125 mg/hour for an average shell containing 2 gm of calcium. It seems unlikely that a rate of absorption of such magnitude from the digestive tract can be maintained throughout the full period of rapid shell calcification, and the evidence suggests that some degree of mobilization of calcium occurs for at least part of the time of shell formation. It has actually been shown that after intravenous injections of parathormone into laying hens, there were significant increases in the excretion of three main components of bone, namely, (1) calcium, derived from the hydroxyapatite crystals, (2) hydroxyproline from the fibrillar collagenous matrix, and (3) uronic acids from the chondroitin sulfates in the amorphous fraction of the matrix. Moreover, the ratios of calcium-hydroxyproline and calcium-uronic acids in the urine increase up to 1 hour. Thereafter, there was a reversal, and the urine became enriched with the organic end products of bone catabolism. After about 2 hours, a new reversal occurred, with signs of a return to the basal state. Parathormone seems thus to cause active resorption of both the mineral and organic phases of bone, but not necessarily simultaneously. The mobilization of calcium from the skeleton of laying hens by parathormone is thus extremely rapid, and this mechanism is obviously capable of providing calcium for eggshell formation within a short time after release of the hormone.

On the other hand, there has evolved in female birds a complex, specific mechanism that facilitates provision of the enhanced calcium requirement of the animal during the period of egg laying. It consists in definite alterations in calcium metabolism elicited by a high output of estrogen by the growing follicle. The female hormone induces a drastic hypercalcemia. Experimentally, blood calcium levels as high as 145 mg% have been obtained in the duck by estrogen injections. The total calcium concentration in the plasma of the laying hen is thus regulated by estrogen as well as by parathormone, but the effects of the two hormones are additive rather than mutually potentiating. Parathormone maintains a suitable level of ionic calcium in the plasma, while estrogen increases total plasma calcium, without increasing ionic calcium, by the formation of a calcium-phosphoprotein complex . This seems due to the large numbers of calcium-binding sites available in the estrogen-dependent yolk proteins as compared with the common plasma proteins (25-fold greater than albumin). The marked estrogen-induced hypercalcemia appears thus as a reflection of an enhanced transport of protein-calcium complex from the liver to the ovary.

Simultaneously, a very specific modification occurs in the skeleton, with extended deposition of a secondary system of highly calcified spongy bone within the marrow cavities, the so-called medullary bone. This medullary bone serves as an advanced, highly labile reserve supply of calcium throughout the laying period. The precise type of interaction of parathormone and estrogen in the control of the metabolism of medullary bone is controversial. Benoit (1950) claimed that in the duck parathyroidectomy inhibits estrogen-induced hypercalcemia, as well as the mineralization of the spongy medullary bone, and brings estrogen-induced hypercalcemia, once obtained, to a dramatic, and finally fatal, decline. However, it has been shown later, in the fowl, that this does not necessarily occur, provided that the birds are massively supplemented with calcium or with vitamin D. Some light may again be thrown on the question if one assumes that estrogen-induced synthesis of yolk proteins requires an adequate concentration of serum calcium ions and that there is an equilibrium between calcium and protein in solution in the plasma, the fine regulation of it being a main function of the parathyroid hormone.

The nature of the control of the resorption of medullary bone, and hence the calcium output from the reservoir, is also still unclear. According to Urist (1967), one striking fact is that parathyroid hormone, which is so active in rapid resorption of metaphyseal and endosteal bone, seems to be without effect on resorption of estrogen-induced intramedullary bone. The fluctuations in rate of deposition or resorption of calcium salts from medullary bone could reflect mainly the corresponding fluctuations of the level of plasma estrogens. On the other hand, as stated above, to others, the parathyroid plays a major role in the control of skeletal resorption during the egg-laying cycle.

Regulation

The chief regulatory role of parathormone synthesis and release undoubtedly is exerted by the feedback action of the plasma calcium level. Thus, high calcium diets (3%) fed to chickens caused, together with a significant hypercalcemia, a marked reduction in the dry, defatted weight of the parathyroid glands, and a correlative decrease in the amino acid uptake by the glands, while low calcium diets (0.3%) led to opposite modifications. Hypophysectomy has also been claimed to lead to moderate depression of the parathyroid glands; hypophysectomy-induced atrophy of the glands is corrected by pituitary autografting. The nature of these relationships, which are probably indirect, is unknown.

Ultimobranchial Bodies

Morphology

The paired ultimobranchial bodies arise from a caudal pharyngeal complex corresponding to the fifth and sixth branchial pouches. Both terminal branchial pouches are essentially evaginations of the posterior wall of the fourth visceral pouch, and in contrast with the first four pouches, they remain small and rudimentary. These primordia develop early into ductless glands in the vicinity of the parathyroid and the thyroid glands. Whereas in adult mammals the ultimobranchial bodies become completely embedded within the thyroid gland, forming therein the parafollicular C cells, both glands remain distinct in most avian species, except in the pigeon, in which a partial incorporation of ultimobranchial tissue within the typical thyroid gland has been claimed to occur. The typical position of the avian ultimobranchial glands (e.g., domestic fowl, turkey) lies bilaterally near the origin of the subclavian and common carotid arteries, and in line with the two parathyroid glands and the thyroid gland, which are located more anteriorly along the carotid artery. In the domestic fowl the ultimobranchial bodies are well delineated rounded organs 1-2 mm in diameter. In laying hens, the weight of a single gland is about 17 mg, as compared with 3 mg for a parathyroid gland and 70 mg for one thyroid gland.

Under the light microscope, the most conspicuous components of the adult ultimobranchial bodies are cystic structures of varying shape and size. Between the cavities, there are cords and sheets of parenchymal cells, mainly composed of light cells in association with a second, scarcer cell type characterized by dense cytoplasm. Numerous blood vessels are present among the cell aggregates. With the electron microscope, the cells bordering the cavities appear columnar to cuboidal, with short microvilli at their luminal pole. Some cells contain large amounts of light granules that give them the aspect of a mucoid cell. More typical are the parenchymal cells of the aggregates that lie between the cavities. Here again, two cell types are present— the dominant light cells that are ovoid to polygonal in shape and contain a pale cytoplasm, and scattered dark cells that are more irregular in shape and have a dense cytoplasm, dilated ergastoplasmic cavities, and numerous and large mitochondria. The light cells contain numerous electron-dense granules which fall into two size groups of about 2000 Å in diameter for the smaller one and about 3000 Å for the larger, and which are lined by a smooth limiting membrane. The ultimobranchial bodies of the chick contain light cells in various states of functional activity. Active cells are characterized by a well-developed rough endoplasmic reticulum, a prominent Golgi complex, and round- to rod-shaped mitochondria throughout the cytoplasm. The fine structure of the light cells strongly resembles that of the light, parafollicular C cells of the mammalian thyroid gland, the granules of which have been shown to have hypocalcemic activity.

The embryonic development of the ultimobranchial bodies of the chick begins with the organization of the compact glandular primordium from the wall of the branchial pouches. From the eleventh day of incubation, ultrastructural indications of secretory activity become evident, with 1100-1600 Å granules appearing within the Golgi complex. The secretory activity increases thereafter with the development of large ergastoplasmic cisternae and increased numbers of granules with diameters of about 2000 Å. Others become as large as 3000 Å before hatching, and even 4500 Å in the newly hatched chick. From the thirteenth day on, some dark cells differentiate; they contain mainly the larger type of granules. After hatching, the principal changes are related (1) to the size of the granules, which decreases to that of the above-mentioned adult types; (2) to the relative number of dark cells, which increases; and (3) to the formation of follicular and cystic structures, which are characteristic of the adult gland and seem to represent degenerative formations.

Hormone

The distinctive anatomical site of the avian ultimobranchial bodies in relation to the thyroid gland has been used extensively in recent years to elucidate the precise site of origin of the hypocalcemic principle that had been previously extracted in mammals from the parathyroid gland or from the thyroid gland. In fact, this hypocalcemic hormone, calcitonin, is the specific hormone of the ultimobranchial gland.

Chemically, calcitonin appears to be a polypeptide. In mammals, the hormone is estimated to contain 32-35 amino-acid residues (32 in swine calcitonin), with a molecular weight of approximately 4000. According to O'Dor et al. (1969), avian calcitonins have molecular weights very similar to the mammalian hormones, ranging between 4300 (domestic fowl) and 4600 (turkey), whereas mammalian calcitonins lie between 3600 (swine) and 4500 (bovine).

Role

Although very little information is available on the physiological significance of calcitonin in birds, its hypocalcemic effect has been clearly demonstrated. Injections of 40 mU/100 gm of calcitonin into 2-month-old cockerels caused a fall of 1.7-2.1 mg/100 ml in plasma calcium in 1 hour. Moreover, injections of extracts of 800 μg of ultimobranchial tissue in rats produced a fall in the serum calcium of 1.17 ± 0.17 mEq/liter associated with a definite hypophosphatemia (-2.95 ± 0.52 mg%). On the other hand, chronic high calcium diet in chickens, with moderate hypercalcemia (increasing from 10.39 ± 0.19 to 13.89 ± 0.39 mg%) led to marked hypertrophy and hyperplasia of the dominant light cells of the ultimobranchial glands. These cells also showed extensive degranulation and ultrastructural modifications that were indicative of increased metabolic and synthesizing activity. Similarly, high calcium diets (3%) that induced hypercalcemia were associated with a significant increase in the dry, lipid-free weight of the chick ultimobranchial gland, and a marked stimulation of amino-acid uptake by the glandular tissue, whereas low calcium diets (0.3%) had the opposite effects.

The high calcitonin content of the avian ultimobranchial glands, together with their high activity (15,000 MCR units/gm extracts for chickens against 250 for hogs, 110 for rats, and 5 for man), the hypertrophy of the glands in laying hens, and the reactivity of their main parenchymal cell type to chronic hypercalcemia, suggest that the ultimobranchial bodies may be involved in regulation of calcium metabolism. Very intense calcium metabolism appears, in fact, as a characteristic of avian species, as it has been shown that birds are able to remove calcium from the blood five times more rapidly than mammals.

However, calcitonin seems to lack any major role in the growth of the skeleton. After removal of the ultimobranchial glands in 1-day-old male chicks. Brown et al. (1969, 1970) were unable to detect any significant difference, at age 1-14-months, compared with sham-operated animals, neither in body weight nor in serum calcium and alkaline phosphatase. Only serum phosphorus was decreased. Furthermore, removal of the ultimobranchial bodies did not alter the amounts of calcium, magnesium, phosphorus, hydroxyproline, and hexosamine in bone, nor did it affect the X-ray density of the tibia or the thickness of cortical bone at 3 months of age. However, the decrease in serum calcium to hypocalcemic levels following parathormone-induced hypercalcemia was prevented. Despite the lack of involvement in skeletal growth and in the bone turnover of the growing chicken, calcitonin seems thus essential in the prevention of hypercalcemia, and may also have a significant role in serum phosphorus homeostasis.

On the other hand, removal of the ultimobranchial bodies from laying hens led to a slight difference in egg size, a trend toward reduced eggshell thickness, a significant decrease in plasma calcium, and trends toward lower plasma phosphorus and lower alkaline-phosphatase activity, which suggest lower rates of bone-calcium turnover in these birds.

Further investigations are needed to understand the mechanism of action of calcitonin. In mammals (rats), the primary effect of calcitonin, like parathormone, appears to be on bone, where it inhibits bone resorption with correlative calcium and phosphorus release and increases the ratio of osteoblasts to osteoclasts.

Regulation

As for parathormone, the main regulation of calcitonin secretion by the ultimobranchial glands depends on the serum-calcium level. This was nicely demonstrated by an experiment with geese in which Bates et al. (1969) measured the rate of calcitonin secretion from a perfused ultimobranchial-parathyroid complex, Perfusion of the complex by hypercalcemic blood caused a net increase in the rate of secretion of the hormone, whereas the removal of the ultimobranchial gland from the tissue perfused with hypercalcemic blood reduced the calcitonin output to zero.

The possible involvement of a neural regulatory mechanism has been recently raised by Hodges and Gould (1969) who have placed emphasis on the rich vagal innervation of the ultimobranchial gland of the domestic fowl. Electron-microscopic studies have demonstrated direct connections between vagal fibers and the densely granulated epithelioid cells of the gland. Moreover, electrical stimulation of the vagus was followed by a very significant decrease of total plasma calcium, an effect that was prevented by atropinization, but could also be initiated by injections of parasympathomimetic drugs.

Thymus

The inclusion of the thymus among the endocrine gland may appear questionable to modern biologists in light of the recently demonstrated role of the thymus in immunobiology. However, the recent isolation of very potent fractions extracted from mammalian thymus, some of which behave like hormones, might lead the way to a reconciliation of the older and newer concepts of the thymus.

Morphology

In birds, the thymus is an elongated lobulated gland that lies on each side along the course of the jugular vein and the vagus nerve, from the third cervical segment downward to the thoracic cavity. There are usually seven thymic lobes on each side.

The embryological origin of the thymus is similar to that of the parathyroid gland in that the third and the fourth branchial pouches are the principal contributors to it. The ventral portion of the third pouch fuses with the dorsal portion of the fourth pouch to form a mass of epithelial cells on each side, the thymus primordium. This primordium elongates then to form an epithelial cord, extending progressively, cranially and caudally, along the jugular vein. The elongated thymus is syncytial and forms a reticular framework as anastomoses of fine protoplasmic threads develop among the stellate epithelial cells. The epithelial cells produce lobules of tissue that progressively give the thymus its typical aspect of irregular broken chains of lobes extending on both sides from the heart to the thoracic region. The thymocytes, i.e., the thymic lymphocytes, appear only secondarily within the syncytial mass of epithelial tissue. The precise origin of the thymocytes still remains an open question. Despite conspicuous evidence pointing to a possible transformation of "undifferentiated" epithelial cells from the primordial thymus into lymphocytes, there are also strong experimental arguments, from thymus grafts, parabionts, and radiation chimeras, that favour the alternate possibility that lymphoid precursor cells, or some type of lymphocytes, invade the thymus epithelial rudiment very early and behave there as typical thymocytes. From the second half of embryonic development on, the thymus lobules have acquired the distinct cortical (dense) and medullary (loose) zones that are characteristic of the adult thymus, the latter zone already bearing rudiments of degenerating Hassal's corpuscules.

As in other vertebrates, the avian thymus increases in size until sexual maturity and then undergoes a marked regression. However, while in mammals the postpuberal regression of the thymus is irreversible, Hohn (1956) has shown that the thymus of a number of birds recovers an enlarged size and a juvenile histological aspect for some weeks, following at least the first sexual cycle. As in mammals, several hormones influence the size and aspect of the bird thymus, although the observed effects may be different. Thus, thyroxine elicits thymic enlargement, while cortisone, corticosterone, and deoxycorticosterone cause atrophy; estrone and testosterone are without effect.

Role

The postnatal avian thymus has a well-known role in the formation of lymphocytes, and in several immunological processes. This also applies to mammals. However, there is now a good deal of evidence that the avian thymus shares with the bursa of Fabricius the immunological function restricted to the thymus in mammals. Thus, neonatal thymectomy in chickens has no consistent effects on immunoglobulin and antibody production, but it diminishes markedly the population of small lymphocytes, which are immunologically competent cells in the blood and in the spleen. By contrast, surgical removal of the bursa of Fabricius at hatching, or inhibition of its development *in ovo* (e.g., by intraallantoic injections of testosterone) is associated with considerable impairment of the capacity to produce immunoglobulin and antibody responses. These findings have suggested the existence of two immunological systems: one, a thymus-dependent system, composed of the thymus and the circulating small lymphocytes that act as recognition agents and produce cell-mediated immune reactions (cellular immunity), such as in delayed hypersensitivity and transplantation immune reactions (coping with homografts); the other, a bursa-dependent system, composed of the bursa of Fabricius, that is responsible for the development of different classes of lymphoid cells that produce immunoglobulins and humoral antibodies (humoral immunity). Both organs direct the differentiation of lymphoid precursor or stern cells into immunologically competent cells capable of reacting to antigens in an appropriate manner within the secondary lymphoid organs (lymph nodules, spleen, etc.).

For many years, oil extracts of the thymus had been claimed to produce lymphocytosis in pigeons. More recently, several mammalian thymic protein fractions with similar actions have been purified; included are the lymphocytosis-stimulating hormone, the competence-inducing hormone and thymosin, a highly purified thymic protein, which acts, when injected, like the two former substances.

A number of other thymic "hormones" have been recently postulated, e.g., an erythrocyte-inhibiting factor; an insulin-like factor; calcitonin, which could be secreted by cells of ultimobranchial origin enclosed within the thymus; thymin, a myolytic substance blocking neuromuscular transmission; promine, a growth-stimulating factor; and a sterilizing factor, which inhibits gametogenesis.

Summarizing, it may be concluded that the main roles of the avian thymus (i.e., lymphocytopoiesis and immunological processes directed against homografts) could conceivably be associated with the production of one or more thymic hormones, like the mammalian thymosin. Further investigation will be needed to ascertain a specific role of the avian thymus in growth promotion and gametogenesis, in accordance with the older observations of Parhon and Cahane (1937, 1939).

Bursa of Fabricius

The bursa of Fabricius is a peculiar avian paracloacal derivative, which has a lymphoepithelial structure closely related to that of the thymus. As stated above, it has been definitely established that the bursa confers immunological competence to birds. More precisely, the bursa is responsible for the maturation of the lymphoid cells, which produce humoral rather than cellular antibodies. There is some evidence to suggest that this prominent effect of the bursa on the development of the bursa-dependent immunological system is mediated through a humoral agent secreted by the bursa.

Extensive research has stressed several relationships between the bursa and other endocrines. It is thus generally admitted that a negative correlation exists between growth of the bursa and development of gonads and of the adrenals. The bursa also has been shown to regress in the presence of exogenous androgens in the Ring-necked Pheasant and in the chick. Similarly, administration of adrenal-cortical hormone, as well as stress, result in involution of the bursa of the domestic fowl, and the pigeon. On the other hand, Riddle and Tange (1928) were unable to detect any change in rate of body growth or onset of sexual maturity in bursectomized pigeons. This was confirmed in the chick by Woodward (1931). However, the data of Taibel (1941) and Click (1955) reveal moderately larger gonads in bursectomized fowl, and bursectomized chicks aged 3-13 weeks had higher body weights than control birds. Neonatal bursectomy also was associated in 5-week-old chickens with enlarged thyroidal follicles and depressed thyroid-^{131}I uptake and with a 50% lowered thyroxine-secretion rate.

A peculiar relationship between the bursa and the adrenal gland has been suggested by some recent investigations. For years it has been found that the ascorbic-acid depletion response of the adrenal following ACTH treatment, which is classic in mammals, did not occur in intact immature fowls. More recently. Freeman (1969a, 1970c) showed that in immature chickens submitted to the stress of handling, adrenal ascorbic-acid depletion occurred very early (10 minutes after the stimulus), whereas ascorbic-acid repletion was completed 1 hour after the initial stimulus despite continued stimulation (handling) at 5 minute intervals. However, ACTH injection superimposed on the handling-stress induced no ascorbic-acid depletion. Indeed, significant increase in ascorbic-acid content of the gland was found; this was interpreted as a result of larger ascorbic-acid uptake than utilization by the gland. On the other hand, the ascorbic-acid depletion response to ACTH was demonstrated, both in adult birds in which the bursa had naturally regressed, or in immature bursectomized chicken, and pigeons. The hyperglycemic response to exogenous ACTH also appears depressed, or at least delayed, in bursectomized chicks, as compared to intact controls. The mechanism of the interrelationships between the bursa and the adrenal glands is still speculative. Freeman (1969a) raised the hypothesis of a loss of adrenal reactivity to ACTH in bursectomized chicks, which could result from the deprivation of some bursal hormone.

Recently, three fractions have been identified chromatographically from saline bursal homogenates of chicks. One of them is assumed to be a steroid derivative.

4

REPRODUCTIVE HORMONES

Reproduction is a cyclic phenomenon, and the majority of avian species are seasonal breeders whose physiological mechanisms regulating gametogenesis are synchronized by environmental stimuli, ensuring that young are produced at the time of year best suited for their survival. Only in some domesticated species, or in those species inhabiting an environment showing little seasonal variation in food availability, has this ancestral cyclic pattern sometimes been lost. Generally, the reproductive organs of most birds undergo a great annual variation in size and functional activity; the whole reproductive process from copulation through the fledging of the young is usually crowded into a few weeks, particularly in those birds that undergo long migrations to high-latitude breeding grounds.

The reproductive system of birds repeats the basic vertebrate pattern. Thus, the seasonal fluctuations in gonadal activity are produced by the environmental stimuli exerting their influence through a central nervous mechanism that leads to gonadotropin release from the adenohypophysis. This is mediated, as appears to be common throughout the vertebrates, by the liberation of neurohumoral substances from the median eminence of the neurohypophysis, these being transported by portal vessels to the pars distalis. The neuroendocrine system, therefore, constitutes a finely integrated and balanced coordinating link between the organism and its environment. The system has evidently had a long evolutionary history, since neurosecretory cells have been identified in primitive coelenterates, platyhelminths, annelids, mollusks, and arthropods, as well as in all the vertebrate classes. Therefore, we have restricted our dissertation to a consideration of the primary sexual organs, the accessory sexual structures, and the associated behavioural aspects that are concerned with the process of reproduction. We have also excluded any extensive consideration of the phenomenon of intersexuality, which is common in certain avian species.

In common with those of all other vertebrate groups, the avian gonads are responsible both for the proliferation of gametes and also for the secretion of the steroid sex hormones that control the development and functional activity of the accessory sexual structures and secondary sexual characteristics. In mammals, these two functions are known to be regulated by the secretion of two distinct gonadotropic hormones from the pars distalis of the anterior pituitary gland, namely, the follicle stimulating hormone (FSH), which is primarily responsible for the regulation of gametogenetic activity of the germinal epithelium, and the luteinizing hormone (LH), which in the male regulates the secretory activity of the interstitial Leydig cells and in the female induces ovulation and the formation of ovarian corpora lutea. In birds, it has sometimes been suggested that the gonads may be regulated by only one type of gonadotropic secretion, but pituitary cytology distinguishes two gonadotropic cell types as in the mammalian situation, and distinct FSH-like and LH-like fractions were earlier obtained by Fraps et al. (1947) from chicken pituitaries using a method previously applied to ovine glands. More recently, more purified samples of these two gonadotropic hormones have been separated from the chicken pituitary gland, and their specificity has been established by biological assay. There is also evidence of two anatomically distinct areas in the hypothalamus of the Japanese Quail responsible for the regulation of secretion of FSH and LH, respectively. Thus, electrolytic or radio frequency lesions in the anterior regions of the infundibular nuclear complex result in regression of the seminiferous tubules without any apparent depression of interstitial cell activity, whereas lesions placed in the medial, ventral, and posterior regions produce, in males, a regression of both gametogenesis and interstitial cell activity, and in females, a cessation of ovulation but no regression of the ovary or oviduct. Similarly, there is good evidence in cockerels, too, that the posterior infundibular region controls the release of an avian LH, whereas the release of FSH is associated with an anatomically separate region.

The male and female gonads are derived from a pair of sexually undifferentiated primordia associated with the intermediate mesoderm (nephrotome). The primordial germ cells within these structures are derived from the embryonic splanchnopleur and migrate in the blood to be housed in these locations and become the presumptive germinal epithelium. They sink below the surface into the connective tissue (stroma). The left presumptive gonad receives a greater compliment of primordial germ cells than the right, and thus establishes an asymmetrical gonadal development which generally persists throughout life. Initially, proliferation of the germinal epithelium in both presumptive gonads in either sex forms a potential testis (medullary tissue). In the female a second proliferation of cells gives rise to a cortex in the left gonad which then becomes the potential ovary. In some species, particularly the domestic hen, a few cortical cords may also sometimes be laid down in the embryonic right gonad as well.

In males, the embryonic gonadal primordia develop into paired testes, but in the female of many species only the left organ develops into a functional ovary, and the right generally remains in an ambisexual state. When the left ovary is removed, or when it becomes non-functional due to some pathological condition, a compensatory development of the rudimentary gonad may take place under the influence of the increased circulation of gonadotropin. In the large majority of cases in which this occurs the rudiment develops into a testis or ovotestis. The experimental induction of this condition in chickens has demonstrated that age can be a modifying factor on the result; full spermatogenesis can develop in the resultant ovotestis if the operation is done at an early age, but if performed in older birds full spermatogenesis is rarely achieved, but ovulations can take place. Under natural conditions, many old domestic hens suffering from senile changes become masculinized, assuming the cock plumage and the capacity to crow, because the rudiment of medullary (testicular) tissue becomes functional. Such sex reversal is also relatively common and often spectacular in pheasant species in which, for example, a somber coloured female Golden Pheasant (*Chrysolophus pictus*) can, as a result of such changes, assume the resplendent plumage of a male. A high incidence of intersexual individuals may also occur in strains of the domestic pigeon, in which apparently a delay in the degeneration of the embryonic cortical tissue causes genetic males to develop with an intact right testis but a left testis in which the cortical component persists and differentiates into ovarian tissue.

Occasionally, adult birds are found with two functional ovaries, particularly among members of the Accipitridae, Falconidae, and Cathartidae, but even in these specimens it is sometimes unaccompanied by a corresponding development of the associated right oviduct. Functional right ovaries are also frequently found in pigeons and in the Herring Gull (*Larus argentatus*).

Individuals without gonads are also sometimes found. Usually these lack both right and left Mullerian ducts and exhibit a general masculine appearance, thus resembling subjects experimentally castrated as embryos. Taber (1964) points out that in man gonadal agenesis or the Turner syndrome is accompanied by loss of one of the sex chromosomes, producing a neuter genotype resembling the female (genetic composition XO which resembles XX). Since males are the homogametic sex in birds, it may be significant that birds lacking gonads resemble the male.

Testis

The essential sex organs in the male bird are the paired testes which, unlike mammals in which they usually become housed in an extraabdominal sac, are located permanently in the body cavity just ventral to the anterior end of the kidneys. Each is attached by a short mesorchium to the dorsal body wall. They are supplied, together with their immediately contiguous ducts, by the spermatic arteries from the dorsal aorta; the testicular veins drain into the vena cava. Because of the asymmetry already established in the presumptive embryonic gonad, the left testis is commonly larger than the right, although in the pigeon and in the turkey, the reverse is frequent.

Each testis is an ovoid, encapsulated body surrounded by a substantial fibrous coat, the tunica albuginea, with a fragile serous outer sheath, the tunica vaginalis. In seasonally breeding species, the testis is subject to great annual variations in size, sometimes by as much as 400- to 500-fold. In the Japanese Quail, for example, testicular weights can be induced to increase from 8 mg to 3000 mg within 3 weeks when birds are artificially exposed to a stimulatory 20 hour photoperiod. Such gross extremes are common in seasonally breeding species and imposes a great strain on the ensheathing tunics, which are replaced annually by a proliferation of new fibroblasts rebuilding a new capsule from beneath the old weakened covering. This upsurge of fibroblasts usually occurs during the postnuptial (regeneration) phase of the testicular cycle, and Marshall (1961a) has suggested that this may be automatically initiated by the collapse of the old tunic. Occasionally, in birds living in captivity or under abnormal climatic conditions, the testes may collapse, with the concomitant formation of a new tunic, even though reproduction has not occurred and the tubules are still packed with spermatozoa. For some weeks after the postnuptial testicular regression, therefore, both the old and new tunics are seen in sectioned material, and the testis appears to be surrounded by a double coat. This transient stage can sometimes be a useful parameter for distinguishing between a juvenile bird, in which the sexual organs have not yet undergone any expansion into a breeding condition, and an adult showing a postnuptial testicular collapse.

Internally, each testis consists of a mass of convoluted seminiferous tubules, which in birds, unlike mammals, are anastomotic and not restricted by septa. They are lined by the germinal epithelium consisting of developing germ cells and nongerminal sustentacular or Sertoli cells. The latter have sometimes been described as forming a syncytium, but electron microscopy reveals them to be separate units whose cytoplasm becomes intimately wrapped around the germ cells, thus giving the appearance of a syncytium when viewed by the light microscope. Lofts (1968) has suggested that this surrounding of germ cells by folds of Sertoli cell might serve to maintain the integrity of a particular population of germ cells in a way analogous to the germinal cysts found in the testes of the anamniotes. During the period of sexual quiescence, the terminal epithelium generally consists of a single layer of stem (type A) spermatogonia and Sertoli cells. Because of the collapsed condition of the tubules, a lumen is sometimes not apparent during this stage because of the consequent compression of the Sertoli cells into the more confined area. However, with the advent of the breeding season, a recrudescence of mitotic activity in the stem spermatogonia causes the propagation of numerous germ cells which mature successively into secondary spermatogonia (type B), primary and secondary spermatocytes, spermatids, and spermatozoa, so that the germinal epithelium becomes several cells thick. Each germinal stage arises in a coordinated fashion, so that the germinal epithelium is not a mass of independently developing germ cells, but is characterized by the different germinal stages being organized into well defined cellular associations. Thus, the sectioned gonad presents an orderly appearance, with the spermatogonia close to the basement membrane and the successive cell types appearing progressively toward the central lumen. There are three clearly defined phases in the rehabilitation of the germinal epithelium to a full breeding condition: (1) the period of spermatogonial multiplication, during which new spermatogonia are formed continuously, and some start maturing into primary spermatocytes; (2) the period of spermatocyte division, when the cells undergo their meiotic division producing secondary spermatocytes which mature into spermatids; (3) the period during which spermatids start elongating and start their transformation into the mature spermatozoa.

In most single-brooded species, particularly those inhabiting temperate and high latitudes, once the gametogenetic resurgence begins, there is a rapid progression into the full breeding condition, and all seminiferous tubules appear uniform in cross section, with innumerable radiating bundles of spermatozoa. This is particularly evident in subarctic and arctic species in which the breeding season is particularly abbreviated and the birds migrate south before the onset of the harsh winter conditions. In continuously breeding forms, however, such as the domestic fowl and feral pigeon, and in multi-brooded species in environments which remain equable for longer periods (e.g., *Columba palumbus*, *Zonotrichia leucophrys nuttalli*, and *Passer domesticus*), the breeding season may be protracted for 3-4 months, and the testes remain spermatogenetically active for a much longer period. In these forms, spermatozoa become released in waves throughout the more protracted breeding season. In mammals, it has been established that the spermatogenetic condition of the germinal epithelium is not synchronous throughout the length of any given tubule but shows a longitudinal progression in the form of a wave of development passing along the length of the tubule. Since the particular stage of development reached in one segment of the tubule may differ from another segment further along, a transverse section of the testis will show an asynchronous pattern, with the germinal epithelium of adjacent sectioned tubules possibly in different stages of the epithelial cycle. Such an asynchronous pattern has sometimes been observed in birds, and Clermont (1958) distinguishes eight distinct stages in the epithelial cycle of the Pekin duck.

There is a tremendous variation in the morphological appearance of avian spermatozoa, but, like those of other vertebrates, they all consist essentially of the sperm head (with an apical cap above the acrosome containing the chromatic material), a midpiece, and a long propulsive tail. Once the spermatozoa are freed from the enfolding Sertoli cells, they migrate from the tubule lumina through the rete tubules and short vasa efferentia into the efferent ducts (vas deferens). These discharge into the expanded distal ends of the ducts, the seminal sacs, from which the spermatozoa are finally ejaculated into the urodeum.

The interstices between the convoluting seminiferous elements are packed with areolar connective tissue containing blood capillaries, lymph spaces, and the interstitial Leydig cells, which produce steroid sex hormone. Large numbers of melanoblasts may also be located in this tissue in some species, imparting a black or gray appearance to the organ. With the seasonal proliferation of the germinal epithelium and consequent expansion of the tubules, the interstitial tissue becomes dispersed and compressed into tight concentrations; fewer interstitial cells appear in testicular sections. Conversely, the interstitial tissue of the regressed testis of a wintering bird appears more conspicuous. These cells undergo well defined cyclic changes in their histophysiology. They show the general characteristics of other steroid-producing tissues. Thus, histochemical techniques have established the presence of the Δ^5-3β-hydroxysteroid dehydrogenase (3β-HSDH) enzyme system in these cells. This enzyme catalyzes the conversion of Δ^5-3β-hydroxysteroids to Δ^5-3-ketosteroids, which is known to be an important stage in the production of such steroids as progesterone and the androgenic steroid androstenedione. It has been found in all types of steroid hormone-producing tissues such as the adrenal, testis, ovary, and placenta, and its presence in the Leydig cells provides strong evidence of their steroid biosynthetic capacity. Ultrastructurally, too, they have the general characteristics attributed to steroid-producing cells, in that, in the breeding season at least, they possess a well-developed agranular endoplasmic reticulum and numerous mitochondria with tubular cristae. Generally, too, lipids are often found in their cytoplasm and react positively to tests for cholesterol.

The interstitial cells appear to be derived from fibroblast-like cells in the intertubular areas. In the young Japanese Quail, Nicholls and Graham (1972) have traced the evolution of the typical mature steroid-secreting Leydig cell from its fibroblast-like progenitor, by means of electron microscopy. It has been clearly established that quail subjected to short photoperiods from hatching exhibit no testicular growth or androgen secretion, but when exposed to a long photoperiod, rapid testicular development and androgen secretion take place. When the interstitial tissue of young quail maintained under a 6 hour photoperiod are examined, three to four layers of spindle-shaped cells with elongate nuclei and prominent nucleoli are observed in the areas between adjacent tubules, with the more centrally located cells being somewhat less elongate but still basically fibroblastoid in their ultrastructural appearance. Generally, granular endoplasmic reticulum is found in these cells, but when the birds are transferred to a highly stimulatory 20 hour photoperiod, the interstitial cells rapidly metamorphose into the typical secretory form; the cells and nuclei expand and develop the characteristic fine structure of steroid-secreting systems with the whole cytoplasmic area becoming filled with smooth endoplasmic reticulum in which are scattered rounded mitochondria with tubular cristae. Nicholls and Graham have recorded the first discernible changes in this metamorphosis at about 3 days after the transfer to the increased photoperiod. Interestingly, Follett and colleagues (1972) have found that plasma levels of LH, measured by means of a radioimmunoassay technique, begin to increase about 4 days after such treatment, which is therefore in good agreement with the ultrastructural evidence.

Histophysiology

General histophysiology

Testicular activity is cyclic, and in some species there is evidence that these cyclic changes may, in part at least, be endogenous. Thus, the testes of young Budgerigars (*Melopsittacus undulatus*), and Zebra Finches (*Poephila guttata castanotis*), become spermatogenetically active and produce spermatozoa even when the birds are kept in almost total darkness. A similar endogenous rhythm has also been recorded in an equatorial African weaver finch (*Quelea quelea*) kept on an unvarying 12 hour photoperiod and thermostatically controlled temperature for 2.5 years. Under these conditions, males show a seasonal testicular development and postnuptial regression similar to that observed in wild populations. In the Pekin duck, too, the testes of birds kept for a

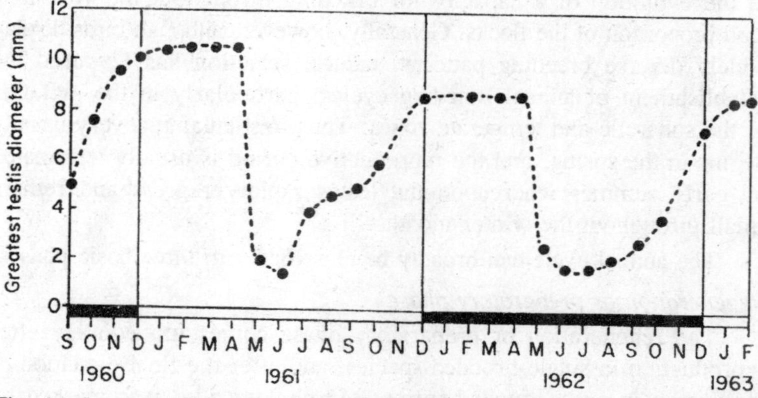

Fig. 4.1. Seasonal changes in the testicular size of the Red-billed Quelea (Quelea quelea) kept under an unchanging daily 12 hour photoperiod.

number of years under conditions of either continuous darkness, or continuous light, similarly undergo a seasonal enlargement and regression.

Marshall regards such an endogenous rhythm as the primary initiator of the seasonal reproductive periodicity. However, in many species such a rhythm is not apparent, and birds like the White-crowned Sparrow (*Zonotrichia leucophrys gambelii*), Brambling (*Fringilla montifringilla*), or Japanese Quail show no spermatogenetic recovery when maintained under winter light conditions and require the seasonal advent of a long photoperiod before a recrudescence of their gametogenetic activity occurs. Even those species possessing an apparently autonomous reproductive rhythm may still be influenced by environmental stimuli, and Marshall (1955) has likened the endogenous cycle to a cogwheel that is seasonally engaged by various environmental "teeth" that produce the final synchronization that ensures reproductive success.

Among various tropical species or species inhabiting a particularly equable environment with a continuous abundance of food, a precise annual breeding cycle may not occur. An equatorial population of *Zonotrichia capensis*, for example, has two complete cycles each year, and Miller (1959) considers this to be an expression of an endogenous 6 month rhythm which is only slightly affected by seasonal rainfall. Even in some temperate zone species, such as the populations of feral pigeons inhabiting our towns, the continuous supply of suitable food provided by a benevolent public and spillage in the docks has resulted in the evolution of a capacity for breeding throughout the year in a high proportion of the flocks. Generally, however, although birds display widely diverse breeding patterns, natural selection has favoured the establishment of annual breeding cycles, particularly in the avifauna of the subarctic and temperate zones. Thus, testicular recovery usually begins in the spring, and the reproductive period is usually terminated by early summer whereupon the testes rapidly regress and remain small throughout the winter months.

The annual cycle can broadly be classified into three basic phases:

Regeneration or preparatory phase

The regeneration or preparatory phase comes immediately after reproduction in single-brooded species, and after the final ovulation of the season in multi-brooded forms. Morphologically, it is marked by the rapid regression of the testes, which can be attributed to a changed and possible diminished gonadotropin release from the anterior pituitary. In many species, this takes place at a time of year when the environmental photoperiod is still highly stimulatory, but the neuroendocrine apparatus no longer responds. The bird is said to be in the refractory period and remains sexually quiescent even when artificially subjected to long photoperiods. Such postnuptial refractoriness is a feature of nearly all species so far submitted to controlled and critical photostimulation experiments, and it has been tacitly assumed or inferred that it is a phenomenon of universal occurrence among avian species. However, recent photoexperimentation on the Wood Pigeon (*Columba palumbus*) has shown this species, and probably others as well, to be without a photorefractory phase. The Wood Pigeon, like the Stock Dove (*Columba oenas*), is multibrooded with a breeding season extending into early autumn where agricultural grain harvests provide abundant food for rearing the later broods. Reproductive activity ceases when autumn day lengths fall below a stimulatory level, and the bird then enters its regeneration phase. Spermatogenesis can, however, be immediately reactivated by artifically prolonging the photoperiod. A refractory period is not therefore a necessity for gonad rehabilitation as previously supposed.

The regeneration phase is characterized by a rehabilitation of both the interstitial tissue and seminiferous tubules, and is a period of sexual quiescence during which the bird displays little or no sexual behaviour. In many species, infiltration of the interstitial tissue by large numbers of leukocytes takes place during the early stages, and the replacement of the weakened testis tunic, and sometimes also the postnuptial molt, occurs during this phase. In the White-crowned Sparrow, for example, a rapid and intensive postnuptial molt is inserted between the breeding period and the onset of autumnal migration. However, the Wood Pigeon molts throughout the breeding season when the gonads are fully active, while the European Turtle-Dove (*Streptopelia turtur*) exhibits a partial molt immediately as it enters the refractory phase but defers the molt of its primaries until it reaches its African wintering grounds. The duration of the regeneration phase varies from species to species, and its termination is probably marked by the completion of the interstitial cell rehabilitation, which may sometimes manifest itself in the form of an autumnal resurgence of sexual behaviour seen in many species. In birds that have a refractory period, the end of the regeneration phase is also heralded by the restoration of photosensitivity.

Acceleration or progressive phase

The acceleration or progressive phase succeeds the regeneration phase and is a period marked by the interstitial cells and seminiferous tubules responding to gonadotropic secretions from the adenohypophysis. The postnuptial molt is usually completed. A recrudescence of gametogenetic activity occurs under favourable environmental conditions but becomes retarded or stimulated by a variety of environmental inhibitors or accelerators that are mutually antagonistic. Thus, the cycle hastens, slows, or sometimes stops altogether, depending upon the factors currently presented to it by the changing environment. Temperature is perhaps the most important modifier of the testicular cycle during this stage, and its effect has been demonstrated experimentally by a number of investigators. In addition, there are many field data that indicate a correlation between temperature and speed of testicular development. Low temperatures will inhibit the cycle of most species, and an unusually cold spring will often nullify the accelerating effects of long sunny days. Among temperate zone birds, some species begin their acceleration phase in late summer or early autumn, but generally this becomes depressed or even halted by the onset of winter conditions until after the winter solstice. In some tropical or xerophilous species, the absence of rainfall may similarly retard the gametogenetic progress during this phase, whereas among waterbirds and many others, a frequent inhibitor is the lack of a safe and traditional nesting site.

The acceleration phase is a period marked by increasingly intensive sexual activity and song, during which gametogenesis leading to the production of spermatozoa occurs. In some species, the winter feeding flocks begin to disintegrate as individuals start territorial selection and defensive displays. The phase varies enormously in duration, both interspecifically and intraspecifically. An interesting example of this is shown by the differences displayed by the resident British population of European Starlings (*Sturnus vulgaris*) and the continental starlings that migrate to Britain for the winter. Spermatogenetic activity begins in the British population in late September, but does not usually progress beyond the proliferation of spermatogonia and occasional primary spermatocytes until February, when a burst of activity populates the seminiferous tubules with secondary spermatocytes and later stages. Continental birds, on the other hand, show no spermatogonial division in autumn, and the first mitoses are not seen until late December or early January, and primary spermatocytes not until early March. Thus, both populations respond to January and February photoperiods, though this marks a beginning of spermatogenesis for continental birds but a resumption of the acceleration phase already started in the British population. Unlike continental birds, the British population shows earlier and more intensive interstitial cell activity in the autumn and winter, which is evident by the earlier changes in bill colouration and development of the vas deferens. This early elevation of androgen titer is also responsible for autumnal sexual displays and even winter breeding in exceptionally mild winters, events virtually absent in the life history of the continental population.

Culmination phase

The culmination phase, during which actual ovulation and insemination occur, may be regarded as a distinct component of the annual cycle, since a species-specific requirement is generally necessary before the cycle can culminate in oviposition. By the end of the preceding phase, the male bird has reached a fully reproductive condition with expanded testes containing seminiferous tubules charged with masses of spermatozoa. Nevertheless, although the internal physiology may be now wholly prepared for reproduction, many complicated behavioural factors may influence the final culmination. The male generally assumes this reproductive state before the female, and the final timing of oviposition then depends on the female receiving the appropriate psychological stimuli, both from her mate and the environment. The appropriate habitat and interpair displays are often essential to stimulate final oocyte development, ovulation, and insemination. The action of stereotyped behaviour in causing specific hormone secretion is now well appreciated, and conditions of captivity, for example, have been shown to inhibit gonadotropin secretion in the Pintail (*Anas acuta*), thus preventing breeding.

Interstitial cells

There is an extensive literature concerning the seasonal changes observable in the avian interstitial tissue, but many of the early reports are contradictory, often being based on unsatisfactory and now outmoded histological procedures. Because of their dispersal by seasonal tubule expansion, the Leydig cells have sometimes been stated to be absent from the gonad at certain times of the year, and an inverse relationship between sexuality and interstitial (Leydig) cell activity has sometimes been claimed. This, of course, is not true and the more recent histochemical and electron microscopic observations have clearly established the close relationship between these cells and the androgen-dependent sexual structures. Furthermore, the selective destruction of the germinal epithelium of cockerels by roentgen radiation, which leaves the interstitium and secondary sexual characters apparently unaffected, also indicate this tissue as the site of androgen production. Tumors of the interstitial cells result in an increased production of androgens and 17-ketosteroids.

In birds, as in the majority of seasonal vertebrates, the interstitial cells undergo well defined seasonal secretory cycles involving a rhythmic accumulation and depletion of cholesterol-positive lipoidal material. The interstitial cells of young birds are generally heavily impregnated with such material. Then, at the approach of the sexual season and consequent buildup of spermatogenetic activity in the seminiferous elements, they become rapidly depleted of their lipids and cholesterol and become strongly fuchsinophilic. In a species such as the Northern Fulmar (*Fulmarus glacialis*), in which the young do not begin breeding until 7 years old, the interstitial cells of newly hatched birds are less lipoidal but become more heavily impregnated when the birds are just over 2 years old. In adults, the interstitial cells of the sexually quiescent winter gonad are generally small and often sparsely lipoidal with numerous fuchsinophilic elements that are more easily seen after dissolution of the lipids in wax-embedded material. During the acceleration phase, these cells rapidly increase in size and there is a buildup of the lipoidal inclusions, so that the interstitial tissue is seen to consist of compressed aggregations of heavily lipoidal and cholesterol-rich cells. They also react positively to tests for 3β- HSDH. Then, as injuveniles, the lipoidal content rapidly diminishes at a time when androgen-dependent sexual displays reach their maximum intensity. During this period, the cholesterol reaction also becomes weaker, and may disappear altogether, although 3β-HSDH activity remains strong. The nuclei of Leydig cells also attain maximum diameter at this time, reflecting an increase in secretory activity. In migratory waders (Charadriiformes), this depletion of interstitial lipids is often evident just before the birds leave their African wintering grounds on their north-bound migration.

With the advent of the regeneration phase, the now exhausted interstitial cells have reached the end of their secretory cycle, and it has been reported that they disintegrate so that ultimately their total number is drastically reduced. It may be that the massive invasion by leukocytes that takes place at this time clears the spent Leydig cells by phagocytic action. However, there is a need for further investigation by electron microscopy to confirm this point, and the possibility that the spent Leydig cells return to an inconspicuous fibroblast-like form should not be excluded. Concomitant with the atrophy of the exhausted generation, a new generation of juvenile interstitial cells begins to arise, presumably by their differentiation from fibroblast-like progenitors, and gradually begin to mature.

This seasonal replacement by new interstitial cells at the end of each breeding period is not unique to avian testicular cycles. A similar phenomenon has also been recorded in some snakes and in the common frog (*Rana temporaria*). Marshall (1961b) suggests that the sequence is part of an endogenous rhythm that can occur even in the absence of any gonadotropic rhythm. Thus, even after complete removal of the adenohypophysis, Coombs and Marshall (1956) have reported that the interstitial cells of domestic cockerels still renew themselves and develop a new generation of Leydig cells with some lipoidal and cholesterol-positive material. It is doubtful, however, whether such cells would ever become secretory in the absence of gonadotropins.

The length of time necessary for interstitial cell rehabilitation probably varies from species to species, but there is evidence that in some birds at least it may be a fairly rapid process. Thus, Lofts and Marshall (1957) have recorded that the newly arisen interstitial cells of fifteen different migratory species were already beginning to manufacture small cholesterol-positive lipid droplets in their cytoplasm when they were autopsied at the time of their departure from Britain on their southward migration.

The cyclical waxing and waning of cellular lipids, which although in itself is insufficient to implicate unequivocally a steroid secreting role, is a useful index of the functional activity of the tissue. The lipids are both cholesterol-positive and strongly birefringent, reactions that are probably indicative of precursor material involved in androgen biosynthesis. The prenuptial buildup in birds often precedes the hypertrophy of the accessory sexual apparatus and behavioural activities thought to be dependent on androgen secretion. In young chicks, for example, the increase in concentration of birefringent material in the interstitial tissue is in close agreement with the gradual hypertrophy of the comb, and in the House Sparrow (*Passer domesticus*) the level of 17α-hydroxylase activity in the interstitial tissue has been shown to increase rapidly between February and March, a time when the interstitial cells are losing the lipoidal material accumulated earlier in January and February.

The sudden depletion of lipids and cholesterol at the height of the breeding activity has also been noted in a number of reptilian species, and it has been suggested that it is probably indicative of a rapid utilization of precursor material at a time of high androgen release.

The observations of Jones (1970) that the height of the epididymidal epithelium in the California Quail (*Lophortyx californicus*) attains its maximum thickness at the time when the interstitial tissue is showing this phenomenon are in agreement with such an interpretation. Certainly 3β-HSDH activity is high at this time. In the Department of Zoology, University of Hong Kong, we have been attempting to correlate these histochemical events with the seasonal fluctuations in androgen production, measured by *in vitro* methods. For this, portions of testicular material have been taken at monthly intervals from the migratory Green-winged Teal (*Anas crecca*) throughout the year and incubated with radioactive pregnenolone as an added precursor. The biosynthetic activity of the tissue has been determined by standard chromatographic analysis of the steroids produced, and the percentage conversion of the added [16-^{3}H] pregnenolone to testosterone per unit weight of tissue has been calculated. Because of the seasonal dispersal of interstitial

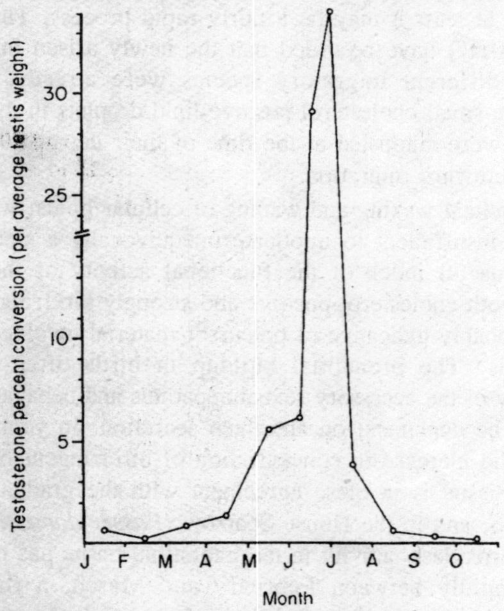

Fig. 4.2. The seasonal fluctuations in the in vitro production of testosterone from radioactive pregnenolone by the testis of the Green-winged Teal (Anas crecca).

tissue during testicular expansion, fewer Leydig cells will be contained in the tissue from the sexually mature testis compared with the same weight of tissue taken from the regressed testes of winter birds. The conversion figures have therefore been multiplied by the mean testicular weight to give a truer picture of the seasonal pattern of steroid production by the whole gonad. They support the hypothesis outlined above. Thus, the gradual elevation in androgen production from mid-February to early June coincided with the appearance of lipoidal inclusions in the interstitial tissue. Lipid depletion was observed in early July at a time when androgen biosynthesis reached a peak, and the subsequent rapid decline marked the bird's entry into the postnuptial refractory period.

Sertoli cells

As well as the interstitial tissue, the Sertoli (sustentacular) cells of the tubule also undergo cyclic seasonal changes involving an accumulation and depletion of cholesterol-positive lipoidal material. Furthermore, ultrastructurally these cells show the fine structure normally associated with a steroid-producing tissue and also react positively to tests for 3β-HSDH. The enzyme 17β-HSDH can also be demonstrated in the tubules of the European Tree Sparrow (*Passer montanus*) at the height of spermatogenetic activity. An endocrine role is perhaps also suggested by the feminization effects seen in the plumage of cockerels suffering from Sertoli cell tumors.

Generally, when the germinal epithelium has advanced beyond the production of primary spermatocytes in the early acceleration phase, there is an absence of lipoidal material from the intratubular components. But as this phase draws toward its termination, and bunched spermatozoa are seen to be radially arranged around the tubule lumen, sections stained with sudan dyes show a light scattering of fine lipoidal granules concentrated in the region of the sperm heads. These small droplets are a component of the residual bodies that form part of the spermatid protoplasm. The residual bodies subsequently become sloughed off from the spermatid during the end stages of spermateleosis and remain behind in the seminiferous tubules after evacuation of the spermatozoa. They are a common feature of the spermatogenetic process of probably all amniotic vertebrates and have been extensively studied in the rat, where they are known to be subsequently phagocytized by the Sertoli cell. In addition to their lipid inclusions, the residual bodies also contain a large amount of RNA, clusters of mitochondria and Golgi remnants, and, in the rat at least, it has been suggested that they may be instrumental in maintaining the radial coordination of the associated germ cell population.

In contrast to the slight scattering of intratubular lipids observed in the fully mature testis, a dramatic metamorphosis into a condition of heavily sudanophilic and strongly cholesterol-positive tubules takes place during the postnuptial regeneration phase. In sectioned material the seminiferous tubules appear to become filled with a dense amorphous mass of lipid and cholesterol, completely occluding the lumen, but with the better resolution provided by the electron microscope, the lipid is generally seen to be within the Sertoli cell cytoplasm. While some of this lipid might be contributed by the lipoidal inclusions of the phagocytized residual bodies, much of it is probably produced by the Sertoli cell cytoplasm. Steroid dehydrogenase activity ceases to be demonstrable in these cells at this time. The tubules remain in this heavily lipoidal state for some time; then, concomitant with the recrudescence of spermatogenetic activity in the adjacent stem spermatogonia, the sudanophilic material rapidly disappears. The duration between the sudden accumulation in the Sertoli cell cytoplasm and the start of lipid depletion varies from species to species and may be as long as several months. For example, in the migratory Whimbrel (*Numenius phaeopus*), the seminiferous tubules have already become heavily lipoidal by the time the animals leave Britain and head south to their African wintering grounds. The gonads remain in this condition for about 5-6 months, and the Sertoli lipid clearance and spermatogenetic recovery begin just before the birds leave the contranuptial area and return north to breed. By the time they arrive in Britain, tubule lipids are absent and testes are reaching sexual maturity. In columbid species, such a massive postnuptial tubule steatogenesis does not occur, and the quantity of Sertoli lipid is very much less.

In view of the fine structure and histochemical evidence, the question arises as to whether this seasonal waxing and waning of cholesterol-rich lipoidal material in the Sertoli cells is indicative of a seasonal endocrine function as is apparent in the adjacent interstitial tissue. Tentative evidence for this has been provided by Lofts and Marshall (1959), who chromatographically analyzed testicular lipids extracted from birds with regressed gonads containing heavily lipoidal tubules but a lipid-free interstitium. The results showed the presence of progesterone, which also correlated with a positive progestogenic reaction in the blood subjected to a parallel bioassay. A similar analysis on birds with gonads with expanded nonlipoidal tubules but heavily lipoidal interstitial cells resulted in an absence of demonstrable progestogenic activity in the blood, and only androgenic steroids were identified in the testis extracts. By present-day standards, the above data would require more rigorous criteria for specific identification of the steroids, but so far as the authors are aware, no further confirmation of these data has yet been attempted. However, there is strong evidence in favour of this hypothesis in mammals and also in reptiles, in which a separation of seminiferous tubules and interstitial tissue has been achieved by microdissection. By incubating seminiferous tubules with labeled precursors, their steroid biosynthetic capacity has been clearly established. Furthermore, in reptiles, Lofts (1972) has shown that testosterone is a major steroid being produced by this tissue, and that the production varies on a seasonal basis.

Control of Testis Function

Effects of exogenous hormones

(a) *Androgen*. There is an extensive literature dealing with the effects of exogenous androgens on the functional activity of the vertebrate testis, and injections of androgenic steroids have been shown to have stimulatory effect in hypophysectomized and intact fishes, reptiles, and mammals. In birds, administration of androgen has sometimes been reported to cause testicular regression, but has also been shown to induce testicular stimulation in adult Ring Doves, European Starlings, Red-billed Queleas, and House Sparrows. Furthermore, in the pigeon, testosterone administration maintains spermatogenetic activity and prevents testicular regression after hypophysectomy and even induces a reestablishment of spermatogenesis in birds in which the gonads have been allowed to regress completely after removal of the pituitary gland. This is in contrast with the situation in the quail, in which Bayle et al. (1970) report that injections of testosterone propionate into hypophysectomized birds fail to prevent the degeneration of the germinal epithelium, although there is evidence that this process is retarded in androgen-treated birds. In *Quelea quelea*, a recovery of spermatogenetic activity can be produced by exogenous testosterone proprionate in the fully regressed gonads of regeneration phase birds, and although Pfeiffer (1947) has reported that spermatogonial multiplication is not augmented in the regressed testes of House Sparrows subjected to similar treatment, it has been our experience that such gonads can be stimulated when sparrows collected in November and maintained under an 8 hour photoperiod are given daily androgen injections. Furthermore, an elevation of endogenous androgens induced by injections of mammalian LH stimulating the interstitial cells has an even greater effect than that produced by exogenous androgens.

Much of the above controversy can probably be attributed to the fact that androgens have been administered to birds without cognizance being taken of the precise reproductive condition of the animal at the time of injection or implantation of the steroid. Thus, androgens given to birds just starting their seasonal testicular enlargement apparently retard the rate of spermatogenetic recrudescence, but if injections are delayed until spermatogenetic development has proceeded beyond the point of primary spermatocyte production, the same androgen treatment can accelerate gonadal enlargement. In the sexually mature *Quelea quelea*, androgen therapy maintains the testes in breeding condition and prevents the postnuptial testicular collapse. The same is true of the European Starling and House Sparrow. In the pigeon, on the other hand, Chu (1940) has reported that daily injections of 2 mg of crystalline testosterone in sesame oil produces testicular atrophy in sexually mature birds. However, this observation may be somewhat speculative, since it was based on the results of only four specimens, and of these one killed after a week was normal and another killed after 30 days still contained spermatozoa. Furthermore, our own observations indicate that daily injection of testosterone propionate into feral pigeons does not cause a testicular regression but, on the contrary, appears to accelerate the maturation of spermatids into spermatozoa. This lends support to the suggestion of Kumaran and Turner (1949b) that androgens facilitate the transformation of secondary spermatocytes in the cockerel, and is also in agreement with its known effects in lower vertebrates.

The stimulatory effects of androgens on the testes of hypophysectomized birds and on the completely regressed testes of intact wild birds is, in both cases, probably a direct effect on the seminiferous tubules, since Lofts (1962b) has shown that the termination of such therapy is immediately followed by a rapid regression. However, although such evidence of a spermatokinetic effect of androgen has sometimes led to the suggestion that a seasonal resumption of interstitial cell activity may be responsible for the spring resurgence of gametogenesis, it does not explain the apparent contradiction that an inhibitory effect is produced during the early acceleration phase under similar experimental conditions. Lofts (1970) has suggested that there may be some antagonism between the exogenous androgen and gonadotropic hormones at a time when the germinal epithelium is

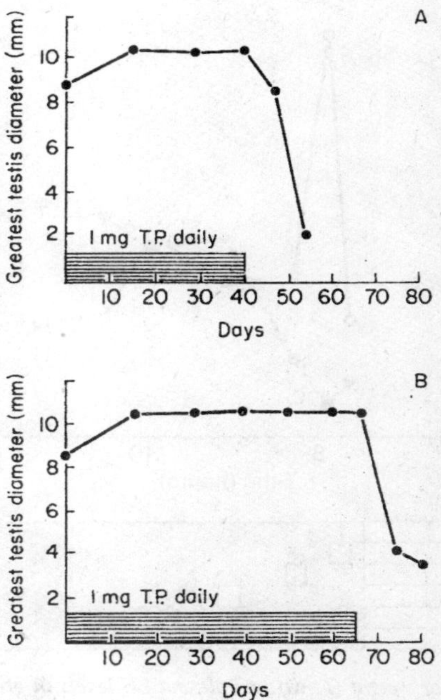

Fig. 4.3. These graphs show the effects on testis size of terminating daily androgen injections in the Red-billed Quelea (Quelea quelea). (A) This experimental bird was kept in full breeding condition by daily injection for 40 days. (B) This experimental bird was maintained in full breeding condition even though the uninjected controls entered and passed through a full refractory phase.

particularly sensitive to FSH, but it is perhaps unwise to extrapolate from experiments using large nonphysiological doses that endogenous androgens have a direct interrelationship with the germinal epithelium. The recent observations by Murton et al., (1969a) on Greenfinches (*Carduelis c. chloris*) throws doubt on such an hypothesis. In an experiment devised to find the photoinducible phase, male Greenfinches with regressed testes were kept in groups under a 6 hour daily photoperiod with an additional 1 hour light pulse given as an interruption in the night. In each group, the night interruption occurred at a different time. After 3 weeks of such treatment, the birds were autopsied for testicular development and the plasma was assayed for LH levels by a radioimmunoassay technique. The results demonstrate that birds of group B had a high plasma titer of LH, which correlated with a highly stimulated interstitium with numerous Leydig cells heavily charged with lipids and swollen prominent nuclei, yet the seminiferous tubules were spermatogenetically inactive. This indicates that, in Greenfinches

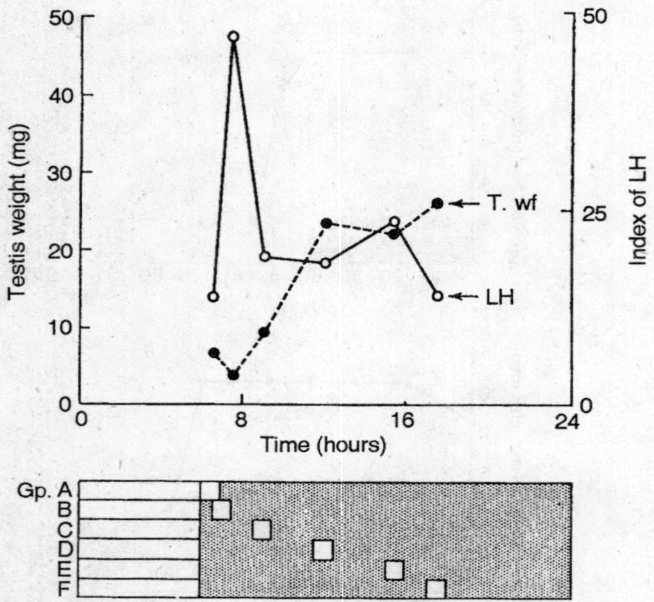

Fig. 4.4. The testicular weight (T. wt) and plasma LH levels in groups of Green-finches (Carduelis c. chloris) subjected to asymmetric skeleton photoperiods.

at least, high endogenous androgen levels produced by the interstitial cells do not stimulate a recrudescence of spermatogenetic activity in the sexually regressed testis, although the possibility of an influence on later germinal stages cannot be excluded. Furthermore, these data do not rule out the possibility that androgens from the Sertoli cells may also be involved.

(b) *Estrogen and progesterone*. There is scattered evidence that the male gonad may produce both estrogens and progesterone under both normal and pathological conditions. Chromatographic and bioassay techniques have indicated the presence of progesterone in testicular extracts of a number of species, and this is supported by the evidence of Fraps and his co-workers, who demonstrated the testicular source of plasma progesterone or a closely allied substance in cockerels. Production of estrogen and progesterone during the normal avian cycle is indicated, too, by the appearance of incubation patches in many species. These are produced by the synergistic action of these hormones with prolactin.

The suppressive effects of exogenous estrogen on the gonadotropic secretion in birds are well established, and estrogen administration to domestic fowl has been shown to produce testicular atrophy involving both seminiferous tubules and an inhibition of androgen secretion with a consequent decline of comb growth.

There is a paucity of information regarding its effects on wild species, but the present authors have recently been investigating the consequence of daily injections of estradiol benzoate on testicular function in pigeons. We find that such injections produce a very characteristic histochemical response; spermatogenetic activity is suppressed, but there is no accompanying buildup of intratubular lipids. In contrast, the interstitial cells become densely lipoidal and distended, with large rounded nuclei. In these cells the cytoplasm is heavily charged with an amorphous mass of sudanophilic material, and they appear as well-marked clusters interspersed throughout an extensive interstitium of small, spindle-shaped juvenile cells containing only a few small cytoplasmic lipid granules.

These results suggest a suppression of FSH leading to an atrophy of germinal stages beyond spermatogonia following estrogen administration. But, in view of the proliferation of juvenile interstitial cells and the massive buildup of sudanophilic lipids in the mature Leydig cells, an increased release of LH is indicated and would be in agreement with the known effects of estrogen in ovulating female mammals. Indeed, recent studies have confirmed that plasma levels of LH are markedly elevated following the treatment of male feral pigeons with estradiol benzoate. It seems likely, too, that there is a direct effect of estrogen on the interstitial cells causing an interference with the later stages of androgen biosynthesis resulting in an accumulation of precursor material, which would also lead to an atrophy of androgen-dependent secondary sexual characters such as comb growth. In the case of feral pigeons, new Leydig cells appear about the middle of the preincubation courtship cycle, and the displays typically occurring at this time are markedly enhanced by exogenous estrogens.

Little is known regarding the effects of progesterone in male birds, and the available data are somewhat contradictory. Farner, on the other hand, mentioned in discussion, that progesterone injections into *Zonotrichia leucophrys* produces a negative result. In pigeons, we have found a variable response to such injections, with some showing no effect on the testes, but others undergoing considerable testicular atrophy. However, it is doubtful whether endogenous progesterone production has any antispermatogenetic effect, since Lehrman (1965) has good evidence that this hormone is released just prior to egg laying, at a time when there is no obvious testicular regression.

(c) *Prolactin*. Prolactin is present in the pituitary gland of male birds and has been known for many years to retard spermatogenesis and lead to testicular collapse in pigeons, cockerels, and some passerine species. Apparently, exogenous prolactin does not induce a testicular regression in *Zonotrichia leucophrys gambelii*, *Carpodacus mexicanus*, *Zonotrichia albicollis*, and *Lophortyx californicus*, but it can inhibit photoperiodically induced gonad growth in *Z. l. pugetensis*, *Z. l. nuttalli*, the Song Sparrow *Melospiza melodia*, the House Sparrow, and Wood Pigeon.

The antigonadal effect of prolactin is suggested to be mediated by a suppression of FSH secretion and produces a buildup of intratubular lipids and cholesterol similar to the naturally occurring seasonal event. In *Ploceus philippinus*, feather regeneration, which is supposedly dependent on LH, is not affected by prolactin administration, but gonadal regression still takes place, again suggesting a suppression of FSH output. However, there is insufficient evidence to attribute a general antigonadal role to endogenous prolactin secretion under natural conditions. In multibrooded species, the seasonal testicular regression does not occur until after the completion of the last clutch, yet prolactin-induced brood patch formation and incubation behaviour take place from the time the first clutch is produced. Furthermore, as in *Z. l. gambelii*, a reduction in gonad size is not produced in prolactin-injected pigeons, although the same treatment is capable of terminating spermatogenetic activity in the closely related Wood Pigeon and has an antigonadal effect in other strains. In the California Quail, Jones (1969c) has recorded the highest pituitary prolactin content in males with testes in full breeding condition, and Gourdji (1970) has similarly recently reported that there is no antigonadal effect in domestic ducks, nor in Japanese Quail, and has further shown that an increase in hypophysial prolactin content under experimental photostimulation is accompanied by an increase in testicular growth. Prolactin does not inhibit the uptake of ^{32}P by the testes of cockerels *in vivo*. The assimilation of this material into the testis is stimulated by gonadotropins, and the measurement of its uptake into the testes of young cockerels is an established bioassay for gonadotropic hormones. Stetson and Erickson found that prolactin administration simultaneously with or up to 18 hours previous to the administration of either purified gonadotropins or pituitary extracts, not only had no inhibitory effect on ^{32}P uptake, but actually acted synergistically at high dose levels.

It seems likely that different levels of tolerance to prolactin secretion have arisen in different species, which would account for the variations in gonadal response. Certainly, there is no noticeable reduction in testis size in pigeons and doves when prolactin-dependent crop gland development and crop milk production is at its maximum, suggesting that columbid species must tolerate high levels of endogenous prolactin throughout the whole of their very protracted breeding seasons; there are, however, histological changes in the testes during the period of prolactin secretion. Hohn (1959, 1962) attributes a similar high testicular insensitivity to prolactin in the Brown-headed Cowbird (*Molothrus ater*) to an evolutionary adaptation to brood parasitism.

Pituitary-gonad axis

In birds, as in mammals, the hypothalamus exerts absolute control over gonadotropic secretion from the adenohypophysis, so that hypophysectomy, pituitary autotransplantation, or sectioning of the pituitary portal vessels produces a rapid testicular regression. The recent separation of avian FSH and LH fractions again shows a similarity to the mammalian situation in that testicular control is mediated by means of a dual hormonal regulation from the adenohypophysis. Lofts and Murton (1968) have proposed that the two gonadotropic hormones might not necessarily be secreted simultaneously, but rather, they may be secreted at different phases. Evidence in support of such a hypothesis appears to be provided by the Greenfinch experiments outlined earlier in this chapter. In these birds, testicular growth (and, therefore, presumably FSH secretion) was most strongly induced in birds (group F) that were given a light pulse 17-18 hours after dawn, whereas the LH peak occurred in response to a light pulse between 7 and 8 hours after dawn. These results imply that there may be two separate hypothalamic centers controlling LH and FSH release, each perhaps receiving information from discrete rhythms of photoperiodic sensitivity, and the results of Stetson (1969, 1971, 1972a,b) provide anatomical evidence of such a mechanism.

Differences in the photoperiodic response mechanism are known to occur in birds, and this has in turn led to a variety of annual breeding patterns. It has been suggested that this phenomenon, in terms of endogenous hormonal controlling mechanisms, must have its basis in differences in pattern in the release of the two gonadotropic hormones, and we have seen in the Greenfinch that a mechanism has evolved in which the photoinducible phase for LH is apparently entirely separate from that promoting FSH release. In *Coturnix*, Follett considers that LH and FSH may be secreted simultaneously at one circadian phase, and this may well also be true in the House Sparrow, where

Murton et al. (1970a) have been able to show that there is a greater overlap of LH and FSH release. In this species, a marked peak of LH release in relation to a photoinducible phase early in the 24 hour cycle does not occur.

Since much of our knowledge about pituitary regulation of avian testicular function has, up to the present time, been derived from studying the effects produced by the administration of mammalian gonadotropins, there is a need for caution and further experimentation with avian preparations before firm conclusions are drawn. However, much of the evidence to date suggests that LH stimulates interstitial cell activity in basically the same way as in mammals. Thus, Leydig cell hyperplasia has been noted in LH-injected young cockerels and was seen to be accompanied by parallel comb growth, and in pigeons, similar LH treatment produces a proliferation in the numbers of recognizable Leydig cells and a buildup in their lipoidal inclusions and steroid dehydrogenase activity. Such results have recently been endorsed by the first preliminary experiments using purified avian LH. Follett et al. (1972) report that daily administration of 50 μg of avian LH for 10 days to sexually regressed young Japanese Quail left under nonstimulatory photoperiods stimulates Leydig cell development into typical secretory form with abundant smooth endoplasmic reticulum.

The fact that exogenous androgens cause a regression when administered to early acceleration phase birds might indicate a negative feedback effect on the anterior pituitary, and the results of Kordon and Gogan (1970) showing that microimplants of testosterone within the ventromedial nucleus of the hypothalamus in ducks induces a testicular regression adds support to such a suggestion. The same investigators have also shown that the threshold of testosterone necessary to block light-stimulated gonad growth varies on a seasonal basis, being low at the time of testicular quiescence, increases when the gonads are stimulated, and decreases at the time of seasonal testicular regression. In *Spizella arborea* and Japanese Quail a similar inhibition of light-stimulated testicular growth is obtained by testosterone implants in the nucleus tuberis region. Evidence of a negative gonadal feedback mechanism is also suggested by the fact that the anterior pituitary of photostimulated castrated birds increase in weight more than the photostimulated intact controls, and that plasma levels of LH become up to five times higher in Japanese Quail subjected to such experimentation. Furthermore, there is a significantly higher acid phosphatase content in the supraoptic and median eminence regions of photostimulated castrated *Zonotrichia leucophrys* than in the intact controls. It would be informative to determine whether exogenous androgen administration would reduce the acid phosphatase content as it does plasma LH in quail.

Injections of purified preparations of mammalian FSH have a stimulatory effect on the germinal epithelium and can restore the spermatogenetic activity of a hypophysectomized bird or an intact bird during the postnuptial regressive stage. Such treatment also produces a depletion of the intratubular lipids, and an inverse relationship seems to exist between the Sertoli cell lipid cycle and the activity of the germinal epithelium. Thus, a termination of spermatogenesis (and presumably FSH output), either naturally during the postnuptial refractory period or artificially as a consequence of hypophysectomy is generally accompanied by an increase in lipids and sterols in the Sertoli cell cytoplasm. Conversely, a recovery of gametogenetic activity during the acceleration phase, or by the exogenous administration of FSH, is always paralleled by a depletion of these substances. Such an association has also been noted in other vertebrate groups, and in view of the steroidogenic capacity of this tissue, it has been suggested that the overall control of spermatogenesis by FSH might be mediated via a regulation of the secretory activity of the Sertoli cells. Such a mechanism would be more in line with the known actions of the other pituitary tropic hormones such as LH and ACTH, which are known to stimulate the secretory activity of other steroid-producing tissues. In the same way, FSH might thus stimulate the secretion of a steroid by the Sertoli cells, which, in turn, would have a spermatokinetic effect on the germinal epithelium. In view of the experimental evidence indicating that exogenous androgens can, under certain situations, stimulate spermatogenesis, an indirect stimulation of the germinal epithelium by FSH acting on the Sertoli cell would appear a more likely hypothesis than one that supposes that a protein hormone, on the one hand, and a steroid, on the other, can cause the same biological action, i.e., a stimulation of spermatogenesis. Evidence in support of such a mechanism is also beginning to appear in mammals and in lower vertebrates.

Refractory period

The development of photorefractivity in those species that have photoperiodically controlled breeding cycles may be regarded as an adaptive mechanism preventing propagation at times when stimulatory day lengths still occur, but when other environmental needs for successful rearing of young are becoming scarce. In migratory species, it causes a premature termination of the breeding period so that the birds are able to undergo their postnuptial molt and autumnal hyperphagia before their contranuptial flight and while an adequate food supply is still available. The length of time before such birds regain their photosensitivity varies, but generally in many temperate zone species the animals remain in this sexually quiescent period for some 3-4 months. By the time they once again become photosensitive it is usually late autumn, and stimulatory light conditions are no longer present in their environment and will not be experienced until the following spring.

The termination of photorefractivity has been shown to be advanced by experimentally subjecting birds to a period of short days, and it is only by exposure to a species-specific term of short photoperiods that these animals retain their capacity to respond to stimulatory day lengths. In some, such as *Junco hyemalis*, *Zonotrichia leucophrys*, *Z. albicollis*, *Sturnus vulgaris*, and *Anas platyrhynchos*, this requirement appears to be absolute, and refractory birds exposed to continuing long daily photoperiods remain sexually regressed. In others, on the other hand, the situation is less rigid, and even when experimentally exposed to continuous summer day lengths, they will spontaneously emerge from their photorefractory state. The domestic duck and *Quelea* are two examples of such a situation. A circadian-based mechanism may be involved in determining the duration of photorefractivity. Thus, when groups of photorefractory House Sparrows are maintained for 35 days under daily photoperiods of 6 hours and an additional 1 hour light pulse given as an interruption of the dark period at 8, 12, or 16 hours from "dawn," respectively, only sparrows receiving a light interruption at 8 hours from "dawn" recover their photosensitivity and show gonadal stimulation when transferred to a stimulatory 16L-8D light schedule. These results indicate that the total daily quantity of light is unimportant, since it was the same in each group. Hamner (1968) similarly claims a circadian component timing the refractory period in *Carpodacus mexicanus*.

The physiological basis for gonadal photorefractivity still remains enigmatic, and the hypothalamus, adenohypophysis, and the gonads themselves have all in turn been suggested as the possible site of the blockage. The early experiment by Benoit et al. (1950) demonstrating that a compensatory hypertrophy of the residual testis occurred in hemicastrated ducks when operated on during the nonrefractory period, but not when the operation was performed during the refractory period, provided evidence suggesting a cessation of gonadotropin output during the refractory phase and indicated that the hypothalamo-pituitary axis and not the pituitary-testis axis as the primary site involved. Furthermore, Stetson and Erickson (1971) have evidence that the gonadotropic content of the pituitary of castrated White-crowned Sparrows shows a spontaneous decline at a time when intact birds enter their refractory phase, again suggesting a hypothalamo-pituitary block independent of any gonadal influence. Nevertheless, it is perhaps still premature to exclude completely any gonadal involvement during this period of photorefractivity. The histological condition of the interstitial tissue of a refractory bird is quite distinct from that of a winter bird that has regained its photosensitive potential but is still sexually regressed, as has been observed in a number of species. This condition can also be induced experimentally. For example, the interstitial tissue of wild Mallards (*Anas platyrhynchos*) caught during their refractory phase in August and maintained in a protracted refractory condition by exposure to artificial summer photoperiods to the end of December has been shown to be much more extensive than that of birds caught at the same time but which have terminated their refractory state while being maintained under an 8 hour winter photoperiod. The testicular interstitial tissue of the latter appears to be much more lipoidal, although spermatogenetically both groups are identical. Hamner (1968) has recorded a similar accumulation of interstitial lipids at the termination of photorefractivity in *Carpodacus mexicanus*, and the same has been noted also in House Sparrows. On the basis of this and other cytological data, Lofts and Murton (1968) have speculated that a release of LH might continue during the refractory phase and be a relevant factor. It is during this time that the differentiation of new Leydig cells occurs, and as has already been noted elsewhere, this rehabilitation of the interstitium with a subsequent release of androgen toward the end of the period is probably responsible for autumnal sexual activity in a variety of species.

Evidence implicating a more positive involvement of the testis in the underlying endocrinological mechanisms of the refractory period is also provided by the recent work of Murton et al. (1970b) on the manipulation of photorefractivity in *Passer domesticus* by circadian light regimes. Although each of three experimental groups of birds were given the same daily ration of light (6 hours plus an additional 1 hour light pulse given as an interruption of the dark period), these investigators found that after 35 days quite distinct cytological differences developed in the testis of each group. Thus, in the two groups that had experienced a light pulse at 12 hours and 16 hours, respectively, from the experimental dawn, the testes had acquired an expanded interstitium with numerous spindle-shaped cells, whereas in the group that had its night interruption 8 hours from dawn the interstitium was less extensive and the Leydig cells remained small but were more heavily lipoidal and the nuclei more rounded. Subsequent exposure to a daily 16 hour stimulatory photoperiod demonstrated that only the latter group had regained their photosensitivity. Thus, the possibility of a feedback effect from the testis during the refractory period cannot be excluded at this time, although a positive conclusion will be reached only when the circulating testosterone levels of the photorefractory bird are eventually measured.

Ovary

Reproduction in birds, unlike the other vertebrate groups, is exclusively oviparous. The left ovary, which becomes the dominant and functional primary sex organ in most species, is attached to the body wall by a short mesovarium in close apposition to the kidneys. It is supplied with blood by the ovarian artery, which is usually a branch of the left renolumbar artery but occasionally is a branch of the dorsal aorta. The organ is drained by large veins that anastomose and converge into an anterior and posterior ovarian vein, both of which empty into the vena cava. Unlike the situation in the male, the functional morphology of the female gonad has been very little studied in wild species, and there is a paucity of information on the seasonal changes in ovarian histology. Most of our data have been derived almost exclusively from investigations of the domestic hen.

In seasonal breeding species, the ovary undergoes the same great variation in size that is evident in the testes, and in the European Starling (*Sturnus vulgaris*), for example, ovarian weight can fluctuate from 8 mg at the time of sexual regression to a weight of over 1400 mg at the height of the breeding season. In the quiescent state, the ovary is a compact, often triangular, flattened organ with small follicles giving the surface a granular appearance. But as it develops into the breeding condition, the follicles become distended and bulge conspicuously from the surface of the ovary, which now assumes the appearance of a bunch of grapes. The largest follicles ultimately become suspended from the surface by a narrow isthmus of tissue, the pedicle. The mature follicle is highly vascular and the surface has a conspicuous scarlike area, the stigma, which delineates the region where the follicle will eventually rupture during ovulation. The stigma appears macroscopically to be avascular, but it does, in fact, have small blood vessels traversing it. In mammals, the ejection of the ovum during ovulation causes a rupture of the follicular capillaries resulting in the blood accumulating in the follicular cavity, but in birds the constriction of spiral vessels prevents such hemorrhage. In domestic species, such as the fowl, the ovary remains more or less in this well-developed state throughout the year (except during molting), but in the seasonally breeding species, the ovary undergoes a postnuptial regression into its quiescent state.

Histologically, the ovary follows the basic vertebrate pattern in consisting of an outer cortex containing the developing follicles surrounding a highly vascular medulla forming the ovarian stroma and composed primarily of connective tissue. At the periphery of the cortex is the germinal epithelium, which in the embryonic stage, starts its gametogenetic activity and proliferates a vast number of primary oocytes. By the time of hatching, the gonad is already endowed with many thousands of oocytes, and no further oocyte formation occurs in adult life. Follicular development proceeds from the endowment of oocytes propagated in the embryonic stage. Of these, only a very small percentage eventually mature fully, and the majority undergo follicular atresia while still in an early stage of maturation. Atretic follicles are, therefore, a prominent feature of the avian ovary.

Unlike the testes of seasonal birds, which in their regressed winter condition are generally gametogenetically inactive and show complete uniformity in histological condition, the ovary always contains large numbers of follicles in various stages of development, as well as a spectrum of corpora atretica in various stages of atresia. In its primary stage the follicle consists of a single layer of flattened granulosa cells ensheathing the oocyte, but as it matures, a multilayered thecal tissue forms around the outside and becomes highly vascularized. During this growth, the granulosa cells become separated by intercellular spaces filled with perivitelline substance, and villus-like elevations arise on the oocyte cell membrane to form the zona radiata. The oocyte completely fills the developing follicle, so that there is no antrum or other internal organization as in the mammalian follicle.

In the Rook (*Corvus frugilegus*), which to date is the only seasonally breeding wild species in which the ovarian cytology and histochemistry has been studied in any detail, stromal fibroblasts become incorporated into the theca interna of the developing follicle and become rounded and glandular as maturation proceeds. With the approach of the breeding season, a proportion of the follicle population progress toward full maturity and eventually enter a phase of very rapid growth that results in conspicuous bulging from the surface. This culminating period of accelerated development is brought about mainly by the deposition of yolk, and Wyburn et al. (1965) have suggested that the ultrastructural elevations of the zona radiata may facilitate the passage of yolk materials into the egg at this time. The final yolk deposition can be very rapid and can, in the pigeon, produce an increase in the maximum follicular diameter from 5.5 mm to 20 mm within 8 days. Sturkie (1954) reports a sixteenfold increase in ovum weight in chickens, in the same period before ovulation. After ovulation, the ruptured follicles rapidly regress, disintegrate, and become cleared by phagocytic action. No persistent corpora lutea of the mammalian pattern are formed, and in the chicken and Rook the postovulatory follicle becomes resorbed within a few days, but in Ring-necked Pheasants (*Phasianus colchicus*) and in the Mallard they persist for as long as 3 months. In these species in which there is a more persistent postovulatory structure, it is derived mainly from the hypertrophic thecal cells.

The numerous follicles that undergo preovulatory degeneration may do so at different stages of maturity. The first observable histochemical indication of this is usually a lipoidal zone surrounding the oocyte, which subsequently spreads so that the whole interior of the follicle becomes occluded with a densely sudanophilic mass of cholesterol-positive material. In the Rook, Marshall and Coombs (1957) report that this fatty atresia is accompanied by a proliferation of fibroblasts that invade the follicle and, for a time, remain as a distinguishable cluster of cells in the ovarian stroma. This phenomenon can also be observed in the Tree Sparrow. These cellular masses subsequently lose their integrity with the disintegration of the follicle and become scattered singly, or in groups, in the stromal tissue. Marshall and Coombs (1957) use the term "exfollicular gland cells" to distinguish these from the stromal interstitial cells that have developed from the ordinary connective tissue cells of the ovarian stroma and are regarded as being homologous with the interstitial Leydig cells of the testis. As in the latter, the ovarian interstitial cells possess the general ultrastructural characteristics of steroidogenic tissue and react positively to histochemical tests for 3β-HSDH, cholesterol, and sudanophilia. Chieffi and Botte (1970) also distinguish a pattern of atresia in which the yolk becomes extruded through the granulosa layer into the surrounding stromal tissue, where it becomes resorbed by phagocytic action. Once yolk resorption is completed only a few vacuolized cells, some capillaries, and strands of connective tissue remain to mark the site of the resorbed oocyte. In other cases, the oocyte may be invaded by cells derived from a hyperplastic activity of the granulosa tissue.

Histophysiology

General histophysiology

Literature on the cyclic changes of the avian ovary is sparse, and very little is known beyond the follicular activity of domestic species. Data on the histophysiological changes in the ovary of the seasonally breeding bird throughout the year are almost nonexistent and have been reported, in any sort of detail, only in one species, the Rook (*Corvus frugilegus*). In this bird, the follicles of the December ovary are too small to be observed on the surface macroscopically, but histologically, sectioned material shows some oocyte development to be taking place internally, and follicles up to 1 mm diameter occur in association with numerous smaller ones. Marshall and Coombs report that the largest follicles at this time of year have distended, sudanophilic, glandular cells incorporated in the theca interna, and that the granulosa layer of some of the biggest follicles disappear as such, as a proliferation of this tissue fills the structure with large lipid-free cells, which later on apparently become secretory. There is relatively little lipoidal atresia at this time of year.

In temperate-zone species, the advent of the spring weather is, as in the male, generally accompanied by a recrudescence of gonadal activity, and developing follicles become increasingly obvious, macroscopically, over the ovarian surface. In *Corvus frugilegus*, follicular recrudescence begins in late January and early February, and from then until the first ovulation in March the whole organ becomes studded with cholesterol-positive lipoidal structures, as increasing numbers of the smaller follicles start developing and become atretic. The stromal interstitial cells also become heavily charged with lipids at this stage and react positively to histochemical tests, both for cholesterol and 3β-HSDH. In the Rook, by March some follicles have enlarged enormously and a few enter the final vitellogenic stage and become ovulated. In a species such as the Band-tailed Pigeon (*Columba fasciata*), on the other hand, no more than one such enlarged follicle is found in the breeding ovary, and this can be correlated with the fact that such species lay only a single egg per clutch. Generally, a species that lays several eggs per clutch will have an ovary containing several large follicles filled with yellow yolk.

After extrusion of the egg, the follicular wall collapses and becomes spotted with lipoidal material. In the hen, the granulosa cells become inflated with lipids and remain conspicuous for about 72 hours after ovulation, but luteinization does not occur and no corpus luteum develops. In seasonal breeders the completion of the last clutch is succeeded by a rapid regression of the gonad, and individual follicles become increasingly difficult to distinguish by the naked eye, though internally great numbers of follicles continue to undergo a slow development followed by lipoidal atresia. According to Marshall and Coombs (1957), a differentiation of a new generation of interstitial tissue, a phenomenon analogous to the testicular interstitial cell rehabilitation during the regeneration phase, takes place in the Rook ovary during the autumn.

There is unequivocal evidence that the avian ovary, like its mammalian counterpart, is a source of androgenic, estrogenic, and also progestogenic secretions, even though a true corpus luteum is lacking. Chromatographic studies have clearly established that this organ, both in its embryonic stage and in the adult bird, has the capacity to produce these hormones, and androgenic, estrogenic, and progestational steroids have all been extracted from the plasma of the domestic fowl. The respective intraovarian sites of biosynthesis of these various steroids, however, is still somewhat speculative. We have already noted that the thecal and granulosa tissue of the preovulatory follicle, the postovulatory follicle, the atretic follicle, the interstitial cells of stromal origin, and also the exfollicular gland cells of Marshall and Coombs, at some stage all contain cholesterol-rich lipoidal inclusions and also have other glandular characteristics that might suggest every one of them to be possible loci of endocrine function, and this is confirmed by the fact that a positive reaction for 3β-HSDH activity has been reported for all these tissues. Furthermore, electron microscopy has shown that all these tissues at some stage possess the fine structure normally attributed to steroidogenic tissue.

Interstitial cells

Because of their apparent homology with the testicular Leydig cells, the ovarian interstitial cells arising from the stromal tissue are generally regarded to be the most likely source of androgenic secretion within the female gonad. Techniques for displaying steroid dehydrogenase activity have indicated that in the hen the steroidogenic potential in this tissue develops early in the embryogenesis of the organ and not just in the post-hatching period, as had previously been supposed, and this has also been shown to be the case in *Coturnix*. In a seasonally breeding bird, such as the Rook, the cyclic development of these cells closely parallels the development of the sexual displays, becoming increasingly prolific and active as sexual activity heightens in the spring and autumn. In the hen, too, the 3β-HSDH activity in these cells increases in intensity as the animal enters its egg-laying period, and reduces in intensity as the follicular phase wanes. The work of Taber (1951) also strongly suggests that a stromal cell is responsible for androgen production. The comb of the domestic fowl is in both sexes under androgenic control and is sometimes used as a bioassay for androgenic analysis. Taber recorded a reappearance of foamy lipoidal interstitial cells concomitant with the decline of androgen production (comb regression) in hens after the cessation of gonadotropin stimulation. It seems likely that in the absence of gonadotropic stimulation and decline in androgen synthesis, the rapid accumulation of cholesterol-rich lipoidal precursor material in these cells resulted in them becoming more prominent. The more recent methods for the visualization of androgenic steroids by means of fluorescent antibody techniques confirm that the interstitial tissue is the primary locus of androgen secretion, though the thecal and granulosa layers also produce a weak reaction.

Boucek and Savard (1970) have recently investigated steroid formation in the hen's gonad at different stages of the ovarian cycle, by a chromatographic analysis of the bioconversion of [^{14}C] acetate into progesterone, androstenedione, testosterone, and estradiol-17β in an *in vitro* incubation. A parallel analysis of the distribution and intensity of 3β-HSDH and 17β-HSDH activity was also carried out to provide data on the intraovarian location of the sites of activity. It was discovered that whereas the same spectrum of radioactive steroids was manufactured, the relative ratios of one steroid to the other varied significantly with the state of the ovarian cycle. Thus, a relatively large proportion of androgenic steroids was synthesized by the gonad of the molting hen, whereas the steroid profile produced by the ovarian tissue of the laying hen showed equivalent amounts of progesterone, androgens, and estrogenic hormones. In the broody hen, chiefly progesterone and estrogen were synthesized with very little androgen. Significantly, the histochemical studies demonstrated that the high androgenic production by the molting hen coincided with an intense 17β-HSDH activity in the stromal cells, and also the thecal tissue, whereas only trace reactions were produced by these cells during the laying stage.

Developing follicle

In mammalian ovarian histophysiology, current opinion generally credits the thecal cells of the developing follicle with the secretion of estrogens, and the granulosa cells with the secretion of progesterone after they metamorphose into the granulosa lutein tissue of the corpus luteum. Support for this latter observation is also provided by the data showing that granulosa cells grown in tissue culture manufacture progesterone. In birds, the evidence is more conflicting, but the tremendous increase in the size of the oviduct which is coincident with sexual maturity is known to be an estrogen-dependent effect and indicates that estrogen secretion appears to be a consistent consequence of follicular development, as is the case in mammals. In a seasonally breeding species, the marked annual fluctuations in oviduct development closely parallel the activity of the ovary. Thus, in *Corvus frugilegus* the oviducal epithelial height in the sexually inactive winter condition is 8 μm, but cellular proliferation begins in early February as the vernal acceleration of follicular development starts, and the epithelial layer rapidly expands to a maximum height of about 40 μm by the end of the month. When follicular development diminishes, the oviducts become constricted again. Estrone, estradiol-17β, and estriol have all been found in avian ovarian tissue, and the former two substances have also been identified in the plasma of the laying hen. In addition to their vitally influential role in the seasonal preparation of the oviduct, estrogens have many effects on a wide variety of somatic and physiological processes in birds.

The seasonal histochemical changes described by Marshall and Coombs (1957) in *Corvus frugilegus* seem to support the hypothesis that the avian thecal cells may similarly be the site of estrogen synthesis as they apparently are in the mammal. Thus, an accumulation of cholesterol-positive lipoidal material builds up in the glandular cells of the theca interna in the largest follicles during the winter months, and then rapidly becomes depleted as estrogen titers are reaching their peak and the mature follicles are undergoing their culminating vitellogenic phase prior to ovulation. By the time yellow yolk appears in the oocytes, only slight traces of sudanophilic material remain in the cytoplasm of the greatly distended thecal cells. Vitellogenesis, and the consequent deposition of large quantities of yolk within the developing oocyte, is an estrogen-dependent phenomenon, and the rapid depletion of the thecal cholesterol and lipids at this stage of the seasonal cycle may be indicative of a rapid utilization of precursor material at a time of high steroid synthesis, as has been suggested to be the case when a similar depletion occurs in the testicular Leydig cells. Certainly, the fine structure of the thecal cells and their positive response to tests for steroid dehydrogenase activity suggest a possible site of steroid synthesis, and the more recent demonstration that isolated thecal tissue from growing follicles of the hen's ovary has the capacity to convert cholesterol to estrogens provides strong support for the suggestion that this tissue may be the locus of estrogen production in the avian ovary. Sayler et al. (1970) have investigated the effects of photostimulation on the distribution of 3β-HSDH in the ovary of young Japanese Quail and have attempted to express changes in steroidogenesis by expressing the intensity of the 3β-HSDH histochemical reaction by means of an arbitrary scale based on the density of the formazan granulation deposited by the different steroidogenic tissues. In *Coturnix* that were raised from hatch to maturity under a stimulatory (16L:8D) light cycle, 3β-HSDH activity increased in all the ovarian steroid producing tissues, but from day 35 onward, the index of 3β-HSDH activity in the thecal tissue increased very rapidly from an average value of 1.3 to 4.3, whereas the enzyme activity in the granulosa tissue was less intense, rising from an average of 1.0 to 2.9 in the same period. Interestingly, the enzyme activity in the thecal tissue of birds kept under a short day length (8L: 16D) was also much higher than that of the granulosa tissue.

According to Chieffi (1967) the granulosa cells are the probable site of estrogen biosynthesis, since 17β-HSDH is limited to this tissue. This enzyme catalyzes the transformation of testosterone into androstenedione and of estradiol into estrone. Arvy and Hadjiisky (1970) have also recorded a greater dehydrogenase activity in the granulosa cells both in the hen and in the quail. However, Marshall and Coombs (1957) point out that the granulosa cells in the seasonally breeding Rook are the only prominent ovarian cells that do not contain cholesterol during follicular development, and electron microscopy studies of the hen ovary have shown that granulosa cells start developing abundant agranular endoplasmic reticulum, mitochondria with tubular cristae, cholesterol, and sudanophilic granules, only at a time when the follicles are ready to ovulate, thus suggesting that their steroidogenic activity has been relatively slight up to this time.

Boucek and Savard (1970) have confirmed Chieffi's earlier observations of the localization of 17β-HSDH in the granulosa cells of the laying hen, but have also shown that in the molting bird the reverse situation develops and 17β-HSDH activity now becomes intense in thecal tissue but absent from granulosa cells, even though *in vitro* incubation and chromatographic analysis of the ovarian tissue show that there is a considerable bioconversion of radioactive acetate into estradiol-17β during this time. The same investigation has also shown that there was little or no progesterone production in these molting hens. In view of the absence of 17β-HSDH in granulosa tissue in these birds, and also the very marked reduction in the intensity of the 3β-HSDH as well, it seems not impossible that the granulosa cells might be involved in the synthesis of progesterone. The ovarian steroid profiles of the laying and broody hens both show considerable progesterone biosynthesis and, in both groups, steroid dehydrogenase activity occurs in the granulosa tissue, being particularly intense in the laying hen. These observations are also supported by the data of Furr (1969a), which show that progesterone occurs in the blood of laying and broody hens, but is not detectable in the blood or ovaries of molting birds. Furthermore, the highest levels of progesterone are found in the follicles and only negligible amounts occur in the ovarian stroma.

Postovulatory follicle

The postovulatory follicle has sometimes been suggested as the source of ovarian progestogenic steroids, and the ruptured follicles have been shown to contain progesterone and have the capacity of synthesizing steroids *in vitro*. Furthermore, the experimental removal of recently ruptured follicles influences the retention time of eggs in the oviduct. Conner and Fraps (1954) have established that there appears to be a quantitative relationship between the amount of postovulatory tissue and oviposition, so that the greater the proportion removed, the greater the retardation in oviposition. It has also been established that when the next-to-last ruptured follicle is removed it has no effect, suggesting that whatever endocrine activity the postovulatory follicle possesses, it is rather short-lived, and this is also indicated by the rapid degeneration that generally occurs in most avian species.

The very transient endocrine phase of postovulatory follicles outlined by the experiments of Fraps and his colleagues is also supported by the electron microscopic studies of Wyburn and his co-workers, who have recorded an extensively developed agranular endoplasmic reticulum in the granulosa cells of the recently ovulated follicle, together with an increase in sudanophilia and cholesterol content. There is also a very intense reaction to tests for 3β-HSDH, but this rapidly diminishes with the degeneration of the postovulatory structures. It seems likely that this very abbreviated postovulatory steroidogenic capacity in these cells is probably a hangover from the activity of the granulosa tissue during the period of high progesterone production immediately preceding the ovulatory process.

Atretic follicle

Cholesterol is abundant in the thecal tissue of the atretic follicles, and during the initial stages of atresia in the granulosa cells. In both layers, a positive 3β-HSDH activity has been reported. The latter investigators also report 17β-HSDH in the granulosa layer. Marshall and Coombs (1957) have suggested that the atretic follicles may be involved in progesterone production, since great numbers of follicles undergo such lipoidal atresia, and, in a seasonal breeder, this phenomenon builds up a considerable reservoir of cholesterol by the time of the seasonal ovulation period. However, Sayler et al. (1970) have established that when young *Coturnix* are reared to maturity under nonstimulatory light conditions (8L:16D), there is an increase in the numbers of atretic follicles, and when the 3β-HSDH reactions in the ovary are examined, sparse granulation is seen in the granulosa cells at the beginning of follicular regression. But in the later stages, this disappeared entirely, suggesting a loss of steroidogenic activity in developing follicles once atresia sets in. It is perhaps premature, however, to conclude that the atretia structures play no significant part in the endocrinology of the avian ovary, and it is evident that there is a need for much more information before the precise locus of any of the ovarian steroids can be established with any certainty.

Vitellogenesis

Attention has already been drawn to the fact that, in the days immediately preceding its ovulation, the avian follicle undergoes a very rapid expansion in size due to the storage of large quantities of yolk in the maturing oocyte (i.e., vitellogenesis). For example, in the chicken, the mass of yolk can increase from 100 mg to as much as 20 gm in the 7 days before ovulation. Although experiments using radioisotopes leave no doubt that protein synthesis can, and does, occur within the growing oocyte, this accounts only for a minor portion of the yolk materials synthesized during this short time interval, and current evidence indicates that the greatest portion of the yolk lipids and proteins laid down in vitellogenesis arise elsewhere in the organism. Basically, the process can be divided into three successive stages, all of which are causally related. These are (1) the synthesis of the yolk proteins and lipids, (2) their transportation to the ovary and uptake by it, and (3) the incorporation of the serum proteins into the yolk elements.

In the avian oocyte, the yolk contains relatively more lipid (33% wet weight) in relation to protein (16%) than is the case in many other vertebrate eggs. In fine structure, it is seen to consist of two major components, namely, the yellow and white yolk granules which appear to be freely suspended in the second component, the yolk fluid. The granules often have discrete limiting membranes and, internally, may contain osmophilic subgranules that are larger in the white granules than they are in the yellow ones. The yellow and white granules are laid down in alternating concentric layers that reflect a diurnal variation in carotenol deposition. Very few mitochondria are observable within the mature yolk mass, except during the early stages of yolk-granule formation, nor is there much evidence of an endothelial reticulum. Much of the protein, together with some of the lipid, largely occurs in the granules, whereas the yolk fluid contains large quantities of lipid globules and a smaller proportion of proteins. Although there are slight differences in their detailed chemistry, the major yolk proteins consist, as in all other groups of oviparous vertebrates, of lipovitellins and phosvitin.

Evidence for the extraoocyte origin of most of these constituents stems from the earlier observations that phosphoprotein appears in the plasma of the laying hen. In view of its chemical similarity with the yolk protein vitellin, it was suggested that plasma phosphoprotein might be a precursor of the latter, which eventually became translocated into the egg. The role of blood proteins in yolk formation was further emphasized by the demonstration, by immunological techniques, that even foreign proteins after being injected into the circulation could be detected in the yolk. Other changes in the constitution of the blood that have been observed to occur as a prelude to vitellogenesis are a very sharp elevation in the level of plasma calcium and also of lipids.

More recently, Schjeide and co-workers (1963) have been able to show that, at the onset of laying, two new serum proteins appear in the hen which are chemically and immunologically similar to the three major egg yolk proteins—phosvitin and α- and β-lipovitellin. The same investigators also confirmed the transference of serum lipovitellin into the egg yolk by means of radioactive tracers. In this experiment serum lipovitellin was first extracted and labeled with ^{14}C, then injected into a laying hen. Within 10 hours, much of the radioactive protein was lodged in the yolk proteins, indicating its transference from the circulation across the egg cell barriers into the yolk. The mechanism of the transmission of such blood proteins into the oocyte is far from clear, but it may be by micropinocytotic activity at the oocyte surface. The follicular epithelium certainly seems to play a significant part, and Patterson et al. (1961) have clearly demonstrated a preferential transfer of serum γ-globulins by the follicular tissue to developing ova and have found that its concentration in the yolk is many times that in the serum. Gonadotropins appear to be important for stimulating the mechanism of this final transference of blood proteins into the oocyte, and in some of the lower vertebrate groups, an increase in serum protein transmission and yolk deposition has been shown to result from FSH treatment, possibly by a stimulation of the process of micropinocytosis.

The liver appears to be the extraoocyte locus of the synthesis of yolk protein and lipid, and hepatectomy has been shown to prevent the typical plasma changes associated with the onset of the vitellogenic phenomenon. Furthermore, radioactive phosphate injected into the wing vein of laying hens has been shown to become incorporated first into a protein-bound component in the liver which then becomes circulated in the plasma as a phosphoprotein, before finally being transferred into the oocyte and localized in the egg yolk. Confirmation is also provided by the demonstration that the *in vitro* incubation of liver slices from laying hens can synthesize phosvitin, which is similar to the yolk phosphoprotein, and Hawkins and Heald (1966) have similarly demonstrated the *in vitro* synthesis of triglycerides and established that it is much greater in laying hens than in the liver of immature pullets. These triglycerides are mainly transported to the oocyte as β-lipoproteins in the plasma, and eventually become concentrated within the fluid yolk compartment of the egg as lipid globules. Although β-lipoproteins occur in the plasma of immature as well as mature birds, there is a pronounced lipemia at the onset of vitellogenesis in the latter.

Control of Ovarian Function

Effects of exogenous hormones

The exogenous administration of different gonadotropic and gonadal hormones and their effects on the female gonad have been less extensively studied than in males, and the little information that is available is generally based on observed changes at the morphological level, without any detailed records of the cytological and histochemical events that might have occurred. Most of our knowledge has been derived from studies on the domestic hen, with particular reference to the effects that such hormone treatment have on ovulation and oviposition. Information about the follicular phase is far less precise.

Androgen

The evidence for ovarian androgen synthesis, and the probable site of production, has already been discussed. The role of androgenic steroids in the regulation of secondary sexual structures and behaviour is also well established, but there is no good evidence that indicates that endogenous ovarian androgens have any significant role in the regulation of ovarian function. The available data are contradictory.

Chu and You (1946) have reported a stimulatory effect on the ovary of both immature and hypophysectomized adult pigeons after androgen administration, and testosterone stimulation of the ovary has also been recorded in *Passer domesticus*. In both these species, androgen injections greatly enhance follicular growth, and Ringoen has noted that the follicular epithelium of treated sparrows becomes stratified. Similarly, stratification of the follicular epithelium and a stimulation of follicular growth have also been noted in the hen ovary after testosterone propionate treatment. An increase in the vacuolation of the cells and a greater ovarian sudanophilia was also noted in the latter investigation. On the basis of observing the effects of a variety of dose levels (0.1-100 μg) of testosterone propionate, Breneman (1955, 1956) has reported that androgen administration increases the height of the follicular epithelium, but a subsequent statistical analysis of his data has led van Tienhoven (1961) to conclude that no dose-response relationship existed and that the differences between controls and androgen-treated pullets could be accounted for by a sampling error. In the American Sparrow Hawk (*Falco sparverius*), large doses of testosterone propionate (140 mg over 30 days) fail to produce any gonadal changes, neither in the left nor right ovary.

Estrogen

Exogenous estrogens are known to delay ovulation in Ring Doves and in domestic hens, and have been reported to inhibit the seasonal development of ova in House Sparrows. Injections into hypophysectomized pigeons failed to induce any significant follicular maturation, and in 30-day-old pullets, estradiol administration did not cause any significant change in ovarian weight, although Breneman (1955, 1956) noted a general increase in sudanophilia and cholesterol content. In contrast to these observations, Phillips (1959) has reported a 32% increase in ovarian weight in 6-week-old pullets injected with diethylstilbestrol, and an even greater ovarian stimulation after similar injections into adult nonbreeding Black Ducks (*Anas rubripes*). The same steroid can also stimulate ovarian development in wild Mallards and is said to act synergistically with gonadotropins to stimulate follicular development. It may be that such contradictory results reflect varying capacities of different steroids to inhibit gonadotropic secretion.

Although to date there is insufficient evidence to attribute any direct effect of estrogenic hormones on gonadal function (i.e., with concomitant pituitary influence), there are ample data to show that such steroids have a vital indirect role by influencing vitellogenesis in the mature ovary. It is now reasonably well established that estrogen stimulates the liver to produce the protein and lipid precursors of yolk. Estrogen injections into female birds lead to a very rapid rise in the plasma levels of fatty acids and lipoproteins, and Roos and Meyer (1961) report that estrogens also boost the serum level of lipoprotein lipase in pheasants. Phosphoproteins also appear in the circulation as a consequence of estrogen administration, and, significantly, the same treatment can even induce similar changes involving the appearance of lipovitellin and phosvitin in the plasma of immature pullets and cockerels.

As might be expected in view of the above observations, estrogen injections also cause parallel changes to take place in the liver, and Schjeide et al. (1963) report that a very large increase in liver volume is produced in domestic fowl as a consequence of estrogen administration. In estrogen-treated birds, the liver tissue undergoes a hyperplasia, and the cells become greatly enlarged with a highly developed endoplasmic reticulum and an increased ribosomal and cytoplasmic protein content. The nuclear RNA content is also doubled in comparison with the average liver cell of nonestrogenized birds. Schjeide (1967) has established that lipid metabolism, and the release of such lipids into the plasma, is greatly stimulated in such birds. Although exogenous estrogen administration brings about a mobilization of yolk precursors, it does not necessarily result in an increase in the number of yolky follicles in the ovary, since gonadotropins are necessary for the transference of these substances from the plasma into the oocyte. Thus, in nonbreeding ducks injected with diethylstilbestrol, only those that were additionally treated with anterior pituitary extract developed large follicles with yellow yolk, whereas treatment with either estrogen or gonadotropin alone failed to stimulate large yolky follicles. The same is true also of wild Mallards.

Progesterone

In domestic fowl, progesterone administration inhibits egg laying if given early in the 24 hour ovulatory cycle, or if given at any time in large doses, and results in an increase in follicular atresia. A similar blockage of ovulation, together with a decrease in ovarian weight and follicular diameter, also occurs in progesterone-injected California Quail when given early in the daily ovulatory cycle. In this latter species the oviduct weight of progesterone-treated females was also significantly smaller than in control birds, suggesting a decline in estrogen secretion presumably caused by an inhibition of FSH release. When, however, progesterone is implanted into immature pullets, it hastens follicular maturation and precipitates precocious egg production, and its administration late in the egg-laying cycle of the adult causes premature ovulation, presumably by stimulating LH release. The experiments of Ralph and Fraps (1960), which demonstrated the ovulation-inducing effects of small quantities of progesterone stereotactically injected into the hypothalamus, but not when injected into the pituitary, indicate that this ovulatory influence is probably mediated via the hypothalamo-hypophyseal axis. The investigations of Kappauf and van Tienhoven (1972) on the progesterone concentrations in peripheral plasma of laying hens have demonstrated that there is a significant rise in plasma concentrations at about 6 hours prior to ovulation.

Prolactin

Although prolactin was the first anterior pituitary hormone to be isolated in pure form, data on its effects on the ovary are much fewer than on the avian male gonad. As in the male, exogenous prolactin injections have sometimes been reported to have an antigonadal effect also in the female, and in the hen, pigeon, and White-crowned Sparrow, treatment with this hormone inhibits follicular development with resultant atresia. In pigeons, hens, House Sparrow, and Bank Swallow (*Riparia riparia*), the ovary has been reported to be regressed during the incubation period, and, in view of the good evidence showing that prolactin release is probably high during this time, such observations are consistent with the experimental evidence showing an antigonadal effect with this hormone. However, in the Rook, the ovaries of incubating females possess many large follicles, and gonadal regression does not appear to take place until after the young have hatched.

Furthermore, Jones (1969a) has reported that prolactin only interrupts, but does not stop egg laying in *Lophortyx californicus* and does not effect follicular diameters or ovarian and oviductal weights. Similarly, Juhn and Harris (1956) found that prolactin blocked the antigonadal effect of progesterone in domestic fowl and only temporarily interrupted egg laying when given alone. Thus, in females, too, there is at present insufficient evidence to attribute a general ovarian regulatory role to prolactin.

Pituitary-gonad axis

Ablation of the anterior pituitary gland produces a regression in the female gonad and extensive follicular atresia. This atresia can be delayed for about 5 days by injections of mammalian gonadotropins and for a longer period by avian pituitary material. Furthermore, ovarian activity and a recovery of ovary-dependent organs can be induced in hypophysectomized hens in which postoperative regression has been allowed to develop, by injection of a crude avian pituitary preparation. In this investigation, Mitchell (1967a) allowed a 10 day

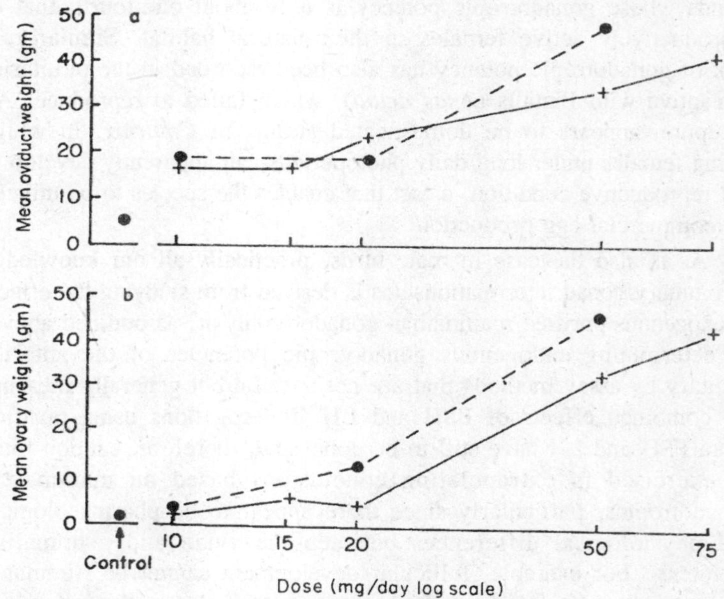

Fig. 4.5. Effects of chicken anterior pituitary extract on (a) oviduct weight and (b) ovarian weight of hypophysectomized hens, in which both organs had been allowed to become completely regressed before commencement of treatment; (●) treatment for 12 days; (+) treatment for 8 days.

postoperative period for complete ovarian and oviduct regression; he then injected different groups of hypophysectomized birds with varying doses of powdered acetone-dried material suspended in isotonic saline.

There is a very close correlation between ovarian function and gonadotropic concentration of the pituitary gland, both in domestic species and in a seasonally breeding bird such as the White-crowned Sparrow, and a similar relationship between pituitary gonadotropin content and ovarian function has also been shown in *Coturnix*, where gonadotropin potency and ovarian development undergo a parallel increase when birds are held under long daily photoperiods from the time of hatch.

Although female birds, like males, show a gonadotropic response to photostimulation, the ovary in most species can rarely be stimulated to full breeding maturity under experimental conditions. A block to growth occurs at the onset of vitellogenesis, presumably because other proximate factors are required. Captivity itself is adequate to stop development, and captive female White-crowned Sparrows have pituitary glands whose gonadotropic potency is only about one-fourth that of reproductively active females in their natural habitat. Similarly, a lack of gonadotropic potency has also been recorded in the pituitaries of captive wild Pintails (*Anas acuta*), which failed to reproduce. An exception appears to be domesticated strains of *Coturnix*, in which young females under long daily photoperiods can apparently develop to full reproductive condition, a fact that enables the species to be utilized for commercial egg production.

As is also the case in male birds, practically all our knowledge on pituitary-gonad interrelationships is derived from studying the effects of exogenous purified mammalian gonadotropins or, as outlined above, by determining endogenous gonadotropic potencies of the anterior pituitary by assay methods that are not specific but generally measure the combined effects of FSH and LH. Investigations using purified avian FSH and LH have still to be done, and, therefore, caution must be exercised in extrapolating conclusions based on mammalian gonadotropins, particularly since there appear to be pharmacological and physiological differences between the avian and mammalian hormones. For example, follicular development cannot be stimulated in the ovaries of immature chicks by mammalian gonadotropins, yet crude avian pituitary extracts produce a marked stimulation. The lack of response to mammalian gonadotropins, however, alters with age, and in pullets nearing sexual maturity, exogenous hormone administration induces follicular development. Dessicated avian pituitary material is also more effective than purified mammalian gonadotropins in causing continued follicle development and ovulation in starved pullets and in stimulating precocious follicular development in the immature ovary.

The differences between gonadotropins of avian origin and mammalian origin is further underlined by the fact that the partially purified avian hormones react poorly in mammalian assays. Interestingly, Licht and Stockell-Hartree (1971) have also tested avian FSH and LH in the lizard *Anolis carolinensis* and find that the relative potency is significantly greater than that estimated from standard rodent bioassays.

Bearing in mind the above reservations, there is good evidence that mammalian gonadotropins have considerable activity when injected into female birds, and ovarian stimulation as a consequence of such treatment has been demonstrated in a number of wild and domestic species. Furthermore, replacement therapy with mammalian gonadotropins is apparently effective in hypophysectomized pigeons. The evidence indicates that, as in mammals, FSH stimulates follicular development and estrogen secretion (as expressed by an increase in oviduct development) in a number of species. For example, the effect of daily injections of 0.1 mg/kg of mammalian FSH for 2 weeks into sexually regressed Green-winged Teal (*Anas crecca*) caught on their wintering grounds in Hong Kong. It can be seen that follicular development has been stimulated and there is a marked hypertrophy of the oviduct, indicating a stimulation of estrogenic secretion. So far, the effects of injecting purified avian FSH have not been recorded, but a partially purified preparation injected into hypophysectomized hens maintained the ovary and oviduct weights at the control level when therapy commenced within an hour after the operation. There was also evidence that yolk deposition also continued after hypophysectomy in these birds. When hormone injections were delayed until 10-15 days after hypophysectomy, however, a resurgence of follicular development could not be induced at the dose level used (0.33, 0.5, and 0.75 mg/day). When avian FSH was tested on lizards, it stimulated a rapid development of the follicles and a significant increase in oviduct weight.

Nalbandov (1959) has pointed out that in birds, unlike most other vertebrate groups in which groups of follicles mature together and are destined to ovulate together, the growth and ovulation of follicles is continuous within the breeding season, and there are mechanisms that must prevent the maturation and ovulation of more than one egg at one time. Thus, the avian controlling mechanisms provide for the existence of a hierarchy of follicles of graded sizes, only one of which becomes ovulated at any one time and usually on any one day. There is considerable evidence that ovulation depends upon a cyclic release of LH, but ovulation will be induced only if the follicle has reached sufficient maturity to respond. Nalbandov (1961) has pointed out that the ovarian follicle is extremely resistant to the ovulatory action of exogenous LH until shortly before the expected release of endogenous LH, and in the domestic hen and also in the Japanese Quail, there is good evidence that plasma LH levels display a significant peak some 6 to 8 hours prior to ovulation. The magnitude of the rise in plasma LH is much smaller than that found in mammals prior to ovulation, and may reflect an avian physiological adaptation, since a massive burst of LH might release more than one ovum from the ovary. Between clutches the level of LH declines and becomes almost undetectable.

The release of LH from the anterior pituitary gland is under nervous control, and lesions placed in the ventral preoptic region of the hypothalamus block the ovulatory process. Such lesions will also prevent progesterone-induced ovulations, indicating that this steroid stimulates LH release via a hypothalamic route, and this is also indicated by the observation that electric stimulation of the preoptic area will induce a premature ovulation

Accessory Sexual Organs

The accessory sexual organs are those parts of the reproductive system exclusive of the testis and ovary that are also vital to reproduction. These comprise the sperm ducts in the male and oviduct in the female, together with any special cloacal differentiation. The reproductive process may also be assisted by other differences between the sexes. For example, plumage colour or pattern may differ, special plumes or wattles may become elaborated in one sex, bill colour may change, or there can be voice differences. Such characters whereby the sexes differ are termed secondary sexual characters, the primary sexual characters (the gonads) obviously being excluded from this definition.

Two pairs of ducts, derived from primitive kidney ducts, develop in both sexes, these being the Mullerian and Wolffian ducts and each pair runs from the gonad to the urogenital sinus. In the male, the Wolffian ducts become the gonoducts draining the testes, and subsequently become the vasa deferentia and epididymides in the adult bird, while in the female they usually remain vestigial. In the female, the left Mullerian duct becomes the oviduct, while the right, and both these ducts in the male, remain vestigial; occasionally the ducts do not atrophy as appropriate. The occasional persistence of the right Mullerian duct in the female appears to be genetically controlled, since a high proportion of certain inbred strains of fowl exhibit two ducts instead of one.

Male

Mature spermatozoa are released into the lumina of the spermatic tubules, which form the highly convoluted network of the testis. From here they are passed by a small number of tubules, the rete tubules, to the highly convoluted vasa efferentia. Outside the breeding season, the rete tubules are inconspicuous, as they are sited in the tunica albuginea of the testis, but they do become enlarged and distinct as two to four tubes extending from the testis with the onset of reproductive activity. In view of the fact that spermatozoa are rarely observed in the rete tubules. Bailey (1953) has concluded that their passage through this region is rapid and probably periodic. The spermatozoa next pass to thin tubules, the vasa efferentia, which also show some seasonal enlargement and which may become secretory. These leave the tunica albuginea and join to form a long coiled tube, the epididymis. This comprises a compact structure closely bound to the testis and lying adjacent to the kidney. With the onset of sexual activity, the epididymis hypertrophies to become a prominent white knob on the side of the

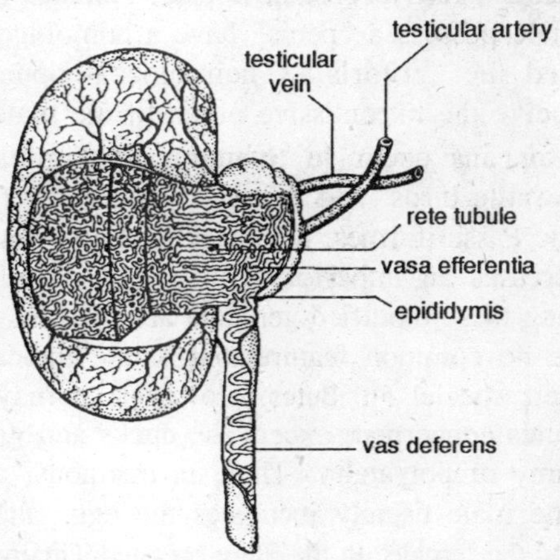

Fig. 4.6. Diagram of the internal network of the testis of the domestic fowl and its relationship with the efferent ducts.

testis. The cells of its epithelium are columnar and ciliated, and secrete a seminal fluid. Sperm are not stored in this organ but progress into a deferent duct, the vas deferens; two muscular tubes one from each of the paired epididymides therefore lead to the urodeum of the cloaca. These vasa deferentia become enlarged, more convoluted, and distinct from the kidney during the breeding season, and before entering the cloaca are expanded to form a seminal sac. This increases thirty-to fortyfold in weight during the breeding season, becoming heavily charged with spermatozoa. A muscular sheath forces the posterior wall of the sperm sacs to expand into the cloaca as erectile papillae and doubtless facilitates the extrusion of sperm during copulation. However, these papillae are minute in most species, and during copulation the cloaca of each sex is everted and the papillae are brought into contact with the opening of the oviduct.

In some species part of the cloaca is modified as a penis-like structure. In waterfowl the penis is a vascularized sac which can be protruded by a muscle and retracted by a ligament. The ligament is bound spirally round the sac and gives the penis a twisted appearance. Sperm move along a spiral groove that passes from the cloacal papillae at the organ's base to its tip. Hohn (1960) has shown that the weight of the penis of the Mallard (*Anas platyrhynchos*) increases sixfold with the onset of the breeding season and that castration prevents such development. Benoit (1936) had earlier noted that at puberty the phallus became enlarged under the influence of testicular hormones, but outside the breeding season it did not regress again to its juvenile dimensions but instead remained fairly constant in size. Females of those species in which the male possess a "penis" have a homologous but reduced structure termed the "clitoris." Their cloacal opening is slightly modified to receive the intromissive organ of the male.

Penis-like organs occur in tinamous (Tinamidae), curassows (Cracidae), the ratite birds, and the ducks and geese (Anseriformes). Alone among the Passeriformes, the Black Buffalo Weaver (*Bubalornis albirostris*) possesses an imperforate penis-like external organ. The groups possessing these modified genitalia are diverse, and the species appear to share no common feature that would indicate an adaptive function for their special attributes. However, it may be significant that all the species concerned, except the ducks and geese, habitually practice polygamy or polyandry. Thus, in tinamous, as in nearly all other ratites, the male usually incubates the eggs and cares for the young unaided by the female. In the Slaty-breasted Tinamou (*Crypturellus boucardi*), the female lays an egg for one male which incubates it and then, in turn, she moves on to lay an egg for another male. Conceivably, each male must be ready to fertilize the female with a reduced opportunity for precopulatory displays that would help synchronize the process, but this view is speculative. The Ostrich (*Struthio camelus*). Greater Rhea (*Rhea americana*), and other ratites, and such tinamous as *Nothocercus bonapartei* perform cooperative laying whereby several females lay their eggs in the same nest that is then attended solely by the male; in some cases the females may repeat the performance for another male. Again selection may be favouring more effective copulatory mechanisms in species that do not have the benefit of a long period of intrapair courtship during which time the endocrine state of the male can become adjusted. The ducks are also unusual in that pairing and copulation usually occur on the water.

As already mentioned, seasonal changes in the size and degree of development of the vasa deferentia, the epididymis, and the penis, if present, parallel the seasonal changes observed in the primary sex organs in those species showing cyclical activity; these structures remain permanently developed in continuous breeding species such as the domestic fowl. This seasonal development apparently depends on androgenic hormones in the case of the vasa deferentia, epididymides, and the seminal sacs, and ablation of the testes will prevent the seasonal development of these organs in seasonally breeding species, whereas injections of testosterone propionate will induce their development into a fully breeding condition. Some reservation is needed in that various pioneer experiments have not been repeated since purified hormones became available, and it is possible that early extracts were contaminated with gonadotropins; this reservation is occasioned by recent experiments into the factors inducing the black bill pigmentation of House Sparrows. Moreover, in the case of the female canary, it seems that estrogenic hormones will induce oviduct development only in subjects kept on stimulatory photoperiods and not birds treated in winter. This suggests that the synergistic action of a gonadotropin is necessary, and it is likely that a parallel situation may apply in the case of the male. Critical experiments with hypophysectomized birds have not been performed.

Maximum development of the epididymides is noted at the beginning of the preincubation behaviour cycle of the male pigeon, and the height of the columnar epithelial cells decreases up to the time of egg laying. If androgen-dependent, this change would indicate a declining androgen titer over the period leading up to egg laying, and for this there is other evidence.

Female

The right oviduct, like the right ovary, remains vestigial except in Raptores, and it is the left that becomes highly differentiated to satisfy the processes involved in producing the complex shelled and pigmented eggs. Even in species in which the incidence of two functionally developed ovaries is high, however, only one oviduct may develop. The developed oviduct is a tube attached to the body wall by dorsal and ventral ligaments and double folds of the peritoneum in which are situated a network of blood vessels; the oviduct is covered by the peritoneum. Its walls embody circular and longitudinal muscle fibers variously developed in different sections, and similarly, there is a variable development of a lining mucous membrane. In places, the membrane comprises longitudinal folds covered with a glandular and ciliated epithelium. Five anatomically distinguishable regions are discernible, namely, the infundibulum, magnum, isthmus, shell gland, and vagina. The extruded ovum is received from the ovary via the open anterior funnel-like infundibulum, which so surrounds the follicle that when the latter ruptures the ovum does not lay free in the body cavity. The ovum is also probably guided into the funnel by the ventral ligament. It is fertilized in the upper thin-walled neck of the oviduct, where the mucous membrane is many-folded and before albumin is added to the yolk. In domestic fowls, fertile eggs can be laid 10-12 days following a single copulation, emphasizing the length of time that sperm can remain viable. However, fertility declines after about 5 days, and it is a fact that wild birds normally copulate many times during the laying period. Polyspermy occurs as several spermatozoids enter the blastodics (12-25 in the pigeon), though only one unites with the female pronucleus. By the time the egg is laid, cell division causes the segmentation area or blastoderm to attain a diameter of nearly 5 mm in the domestic fowl.

The middle portion of the oviduct is the highly glandular magnum, where the white albumin is formed; this is deposited round the ovum in four layers to make a complex structure. In the hen, this process takes from 3 to 4 hours to complete. The histological development of the oviduct has been studied in fowls and in a careful investigation of the domestic canary. While the ovaries comprise up to 1.5% of body weight of the canary, there is an approximately linear relationship between oviduct and ovary weights (both represented as percentages of total body weight), as might be suspected since oviduct development depends partly on ovarian hormones. The formation of albumin coincides with the enlargement of one ovarian follicle more than the rest, and this also correlates with the completion of defeathering in the course of brood-patch development and intensive nest-building behaviour. Relatively little oviduct development is apparent until nest-building activity begins; indeed only the division of epithelial cells of the magnum is to be noted prior to the collection of the first nest material. However, within the period of active nest-building, the oviduct increases markedly in length (decreasing again after egg laying), the epithelium divides and invaginates to form tubular glands, and these become distended and their cytoplasm charged with albumin granules; at the same time, the mucosa becomes very folded. The epithelium differentiates during this time into ciliated columnar cells and goblet cells. The histology of the magnum apparently remains uniform along its length at all stages of development.

After albumin has been deposited, the egg next passes into a less glandular and more muscular isthmus of the oviduct where the two shell membranes are secreted, these being comprised of felted protein fibers cemented together with albumin. The outermost membrane has a rough texture and so provides a key for the calcarious shell. This last, together with pigment, is added in the wider and expandable uterus, or shell gland. The shell is secreted in the form of calcium salts (about 97%) supported on an organic matrix of protein fibers. Any ground colour is deposited during the final stages of shell formation, while blotches, streaks, or other surface markings are acquired after the shell has been completed. Because the egg is rotated on its long axis surface markings are often given a spiral configuration. The rotation is achieved by a powerful sphincter muscle, which when relaxed, allows the egg to pass into the cloaca and to be laid pointed end first. In the domestic fowl, about one-third of the eggs become reversed; the vagina bulges beyond the sphincter thereby turning the egg over, whereupon it is eventually laid blunt end first; whether this happens in other birds remains uncertain. It appears that this eventual oviposition is effected by a muscular contraction of the shell gland (uterus) regulated by the secretion of posterior pituitary hormones. Thus, a sudden decrease in the content of vasotocin in the neurohypophysis takes place in domestic hens and is correlated with a rise in plasma levels, coincident with oviposition. Furthermore, a premature expulsion of the egg can be induced by exogenous oxytocin and vasopressin and also vasotocin, the latter being the most effective.

The seasonal hypertrophy of the oviduct that occurs with sexual recrudescence (gonadotropins are implicated) is dependent on estrogen secretion by the ovary and estrogenic hormones have been shown to cause marked morphological changes in immature chick oviducts and to stimulate the differentiation of mature cell types. But other hormones are involved in the full development of the tract, for progesterone and prolactin augment the increase in weight of the oviduct following estrogen injection in Ring Doves and canaries. In domestic fowls, estrogen alone will cause growth of the albumin-secreting glands and the secretion of ovalbumin. Daily injections (5 mg) of diethylstilbestrol into 4-day-old chicks causes a 300-fold increase in ovalbumin production within 15 days, and an increase from 5-10 mg to 1.5-3 gm in the oviduct weight is recorded after 3 weeks of such estrogen treatment. Progesterone administration, on the other hand, appears to stimulate only the synthesis of the egg-white protein avidin, and this has been shown to be effective, not only *in vivo*, but also *in vitro* when minced chicken oviduct tissue is incubated with progesterone. At the cellular level, O'Malley and his colleagues have demonstrated that estrogen treatment stimulates a large increase in nuclear transfer RNA and a definite but smaller increase in cytoplasmic transfer RNA.

Secondary Sexual Characters

Darwin appreciated that the physical differences between sexes in plumage and voice or their possession of various adornments were associated with courtship and that sexual selection depends on the success of certain individuals over those of the same sex in terms of producing surviving progeny. Because females normally choose their partner, and frequently rely on visual cues, isolating signal characters have evolved in many male birds to prevent confusion with, and hence hybridization between, closely related species. The avoidance of such interspecific competition for mates is to be distinguished from intraspecific competition for a mate, which, as Huxley (1938) recognized, leads to the maximum elaboration of display plumage in polygamous species.

Where a permanent plumage difference between the sexes imposes no other disadvantage, sexual dimorphism may have a genetic basis, so that hormones are unable to change the inherited sex type. Examples are provided by the House Sparrow (*Passer domesticus*) and European Bullfinch (*Pyrrhula pyrrhula*).

Witschi (1961) has provided examples of the endocrine basis of various secondary sexual plumage changes and acquisition of other nuptial adornments. Several of these variations have served for the bioassay of hormones. The development of the comb and wattles of newly hatched chicks, irrespective of sex, depends on androgen and can be stimulated by injections of exogenous testosterone. The pigmentation of the bill of birds is of two main kinds: (1) that produced by deposition of brown and black melanins in melanophores that become injected into those epidermal cells moving out from the area of proliferation and (2) yellow, red, and orange carotenoids that are not synthesized by the bird but are absorbed from the diet. The European Starling has a black bill during the contranuptial season, while both sexes acquire a yellow bill with onset of the breeding season. Injections of androgens induce the yellow colour, but estrogen and progesterone do not; castration causes the loss of the yellow colour. In *Quelea quelea*, the bills of both sexes are red during the contranuptial season, but in females this colour changes to a straw colour during the breeding season and is an estrogen-dependent response.

The bill of the House Sparrow (*Passer domesticus*) turns black during the breeding season under the supposed influence of androgen; castration results in a loss of pigmentation in subjects with black bills. Testosterone propionate, however, will not induce a black pigmentation in sparrows captured in winter and held under nonstimulatory day lengths, suggesting the implication of gonadotropins in the response. Exogenous LH plus FSH or androgen plus FSH will induce a black bill pigmentation in subjects kept on short day lengths so that endogenous gonadotropin secretion is inhibited. Colour change in the beak of the Paradise Whydah (*Vidua paradisaea*) apparently depends on LH and not gonadal hormones.

The assumption of secondary sexual plumage seems in many cases to depend on steroid hormones; for instance, castrated Ruffs fail to don the pectoral display plumes described above. Similarly, castrated male Black-headed Gulls (*Larus ridibundus*) fail to acquire the brown head typical of the breeding season. In phalaropes the female dons the more colourful breeding plumage, and in both Wilson's Phalarope (*Phalaropus tricolor*) and the Red-necked Phalarope (*Phalaropus lobatus*), it has been shown that the assumption of this plumage is dependent upon androgenic steroids. Thus, administration of testosterone propionate (but not estradiol or prolactin) to birds of either sex will stimulate the growth of feathers of the nuptial type in plucked areas. The reason that this colourful plumage is normally shown by the female and not the male can be attributed to the fact that there is a higher androgen secretion in the female, the reverse of the situation in nearly all other species. In contrast to the mechanism in the preceding species, in which the acquisition of the nuptial plumage is apparently determined by the elaboration of gonadal androgens, the male-type plumage of the weaver *Euplectes orix* has long been known to depend on a gonadotropin, supposedly LH, and the species has long served as the bioassay animal for this hormone. In this, and other weaver birds, a castrated male continues to develop a nuptial plumage. Females do not normally develop a gaudy plumage, but will do so after ovariectomy. In the natural condition ovarian estrogens inhibit potentiality of LH for stimulating nuptial plumage, and this is confirmed by the injection of estrogen into male weavers, which inhibits their assumption of the nuptial dress. Exogenous estrogens have a similar effect when injected into male fowls. Many of these experiments, however, were performed before purified hormones became available. Furthermore, because subjects were often held under photostimulatory day lengths, some reservation is needed in accepting previous conclusions about many of the supposed hormonal mechanisms governing secondary sexual characters.

Incubation Patches

In many avian species (but not all) the ventral apterium becomes defeathered and highly vascular and edematous before or during egg laying. Hyperplasia of the epidermis also occurs, producing up to a sixfold increase in thickness. These so-called incubation patches, or brood patches, are in close contact with the eggs during incubation and supply the necessary warmth for development of the embryos. In addition, such a structure provides a sensitive surface in contact with the nest and eggs, which in the domestic canary at least, is a source of tactile stimuli influencing nest-building behaviour. The sensitivity increases as the patch develops.

Incubation patches usually develop in the females only (e.g., House Sparrow, White-crowned Sparrow), but in some species they may occur in both sexes (e.g., starlings and swallows). In the Red-necked Phalarope and Wilson's Phalarope, they develop only in the males, which in these species incubate the eggs, the female taking no part in this activity. Generally, there seems to be little correlation between the incubation habits of male birds and the possession of a brood patch, since the incubating males of some species, such as the House Sparrow and Bushtit (*Psaltriparus minimus*), lack it, whereas the nonincubating cocks of others, such as the American genus of flycatchers (*Empidonax*), possess it.

The chronology of the morphological changes that constitute the formation of an incubation patch appears to differ in various species. In the passerine species, defeathering occurs before egg laying, but the timing of patch development in galliformes appears to be more retarded, so that defeathering becomes marked in Ring-necked Pheasants only after the laying of the fourth egg, and in California Quail after the tenth egg. In *Zonotrichia leucophrys*, Bailey (1952) reports that defeathering from the ventral apterium is completed several days before oviposition, and increased vascularization and edema is not apparent until completion of this stage, but in the Red-winged Blackbird (*Agelaius phoeniceus*), Bank Swallow, canary, and House Sparrow, epidermal hyperplasia, increased vascularization, and mild edema occur before defeathering is complete and, in some birds, begins weeks before egg laying. Thus, in both the European Starling and House Sparrow an appreciable increase in the numbers of dermal blood vessels is noticeable a month before oviposition. The increase in vessel diameter, however, occurs mainly during the egg-laying period.

The development of the incubation patch is under endocrine control and involves hormones of both gonadal and adenohypophyseal origin. Generally, it is formed under the influence of estrogen and prolactin, which are synergistic in their effects. In the passerines, exogenous estrogen administration has been shown to induce increased vascularization and defeathering in both intact and ovariectomized birds, but similar treatment of hypophysectomized birds produces vascularization only. When estradiol in combination with prolactin is given to hypophysectomized specimens, the complete patch is developed; in hypophysectomized *Zonotrichia* prolactin alone has no effect. Since the effect of estrogen on defeathering is augmented by prolactin, and the combination of these two hormones is necessary to produce full development in hypophysectomized birds, it seems likely that, in female passerines at least, this phenomenon is normally dependent upon the endogenous secretion of both these hormones. In California Quail, however, prolactin appears to play a more significant role than in the passerine mechanism, since Jones (1969b) reports that in this species injection of estradiol into reproductively active females fails to have any significant effect. Prolactin alone, on the other hand, stimulates epidermal hyperplasia, and when given in combination with estrogen produces full incubation patch development (i.e., defeathering, edema, vascularization, and dermal thickening). In this species, and also in the Bobwhite Quail (*Colinus virginianus*), prolactin-induced hyperplasia of the incubation patch can be stimulated not only *in vivo* but *in vitro* as well. It seems that only the ventral skin is responsive to hormonal effects, since Jones et al. (1970) have shown that estrogen plus prolactin induces the ventral abdominal skin of the chicken to undergo hypervascularization and epidermal hyperplasia only if the skin is *in situ*. Ventral skin transplanted to the back exhibited epidermal hyperplasia but not hypervascularization. Dorsal skin was not responsive *in situ* nor when transplanted to the ventrurn. Thus, hypervascularization of the chicken incubation patch is site-specific, but the epidermal hyperplasia is not. Furthermore, the data show that epidermal hyperplasia is not dependent on hypervascularization.

Progesterone has been found to augment the effects of estrogen in inducing defeathering in the female canary but is not essential for this process, since it can be induced by estrogen alone in ovariectomized birds. When progesterone is given alone it has no effect on vascularization in canaries, House Sparrows, or California Quail, and in the last species it is also less effective as a synergist with estrogen. In the canary, this hormone has been found to be necessary for the characteristic increase in sensitivity of the ventral skin during development of the incubation patch.

In *Phalaropus lobatus*, *P. tricolor*, and *Lophortyx californicus*, in which the males develop incubation patches, it has been shown that exogenous testosterone plus prolactin can induce incubation-patch development. In the natural situation it seems likely that in such species, vascularization occurs when circulating androgen levels are high, and before defeathering of the patches is completed, the latter event being delayed until circulating prolactin levels increase, probably at the time of testicular regression. In the passerine species in which males usually do not incubate, exogenous androgen plus prolactin do not cause such feather loss in the cock, and in the parasitic Brown-headed Cowbird (*Molothrus ater*), a species that does not incubate its own eggs or develop a brood patch, the ventral skin of either sex is insensitive to estrogen-progesterone injections.

In the Red-winged Blackbird, it has been shown that the experimentally induced patch (by exogenous estrogen or estrogen plus prolactin) differs histologically from those of wild incubating birds, and in the House Sparrow, Selander and Yang (1966) have shown that the normal patch is more edematous than the hormonally induced ones, and it may be that the sensitivity to tactile stimulation that has been demonstrated by Hinde and his colleagues may be necessary for the complete development of the patch.

Breeding Behaviour (Endocrinological Basis)

As already made clear in this chapter, gametogenesis depends on a changing pattern of hormone production culminating in the animal entering the phase of overt reproductive activity. At this stage, various behaviours become manifest to assist the process of breeding. These include the acquisition of a territory and its advertisement by song, the attraction and pairing with a mate, and the initiation of displays that lead to egg laying and nestling production. Increasing titers of androgenic or estrogenic hormones have long been assumed to be responsible for many sexual behaviours in male and female, respectively, but it is becoming increasingly apparent that the endocrine basis of reproductive behaviour is complex and often highly specific. Research has been concentrated on various pigeon species and the canary, and while these are possibly reasonably representative of the class Aves, cognizance must be made of the possibility that other species have evolved special endocrine adaptations to suit their own peculiar ecological requirements.

Unpaired, yet sexually mature, male feral pigeons will usually advance with the bowing display toward other members of the same species. Sometimes, an otherwise responsive male will fail to react to a receptive female but respond to other females, and it appears that the birds can recognize particular plumage morphs with which they are reluctant to pair. In free-living populations, this results in certain pairings of like individuals occurring with less than expected frequency, and the aversion appears to function as an outbreeding mechanism preventing the production of homozygous birds that have a reduced fertility; the gene concerned improves fertility and lengthens the breeding season if present in the heterozygous condition. In Ring Doves, much variation in the range of displays exhibited by males during the initial courtship phase could be attributed to their previous experience; an experienced bird might reduce or omit certain of the more aggressive postures. Previous breeding experience by either male or female contributes to and improves the reproductive performance of Ring Doves under captive conditions. Fundamental endocrine-behavioural interactions of this kind doubtless underlie the general fact that under field conditions older and more experienced birds reproduce more successfully.

There is good evidence that androgens mediate bowing behaviour in both male Ring Doves and also feral pigeons, and they restore male-type behaviour to castrates, increase social status and aggressiveness, and induce malelike behaviour in females. In the phalaropes, in which there appears to be a reversal of the normal situation and females display more aggressive patterns than the male, Hohn and Cheng (1967) have shown an unusually high ovarian testosterone content. Both Goodwin (1967) and Davies (1970) have shown that the bowing display differs from species to species, exhibiting a typical intensity in each. Davies (1970) who also examined F_1 hybrids of *Streptopelia* doves showed that these displayed with a characteristic bowing pattern that depended on inherited components of the display. In particular, the rate of bowing proved to be species-specific, while components of the bowing vocalization were similarly inherited in the same manner as morphological characters.

Bowing appears to function as a ritualized appeasement display; it combines crouched submissive postures (at termination of downward bow) with tail-fanning and head-lowering reminiscent of and probably derived from braking movements. However, the upright and advancing components of the display sequence embodying an almost "goose-stepping" gait represent aggressive components. The ambivalent nature of the display is indicated by the fact that if a bowing male catches up with the subject of his attention, he may (1) peck aggressively, (2) mount and attempt to copulate, or (3) turn and flee. Thus, the display involves causal factors for attacking, fleeing, and mating, and it is significant that different exogenous hormones injected into males at the beginning of their preincubation behaviour cycle will alter the relative importance of these various display components. Exogenous FSH increases the incidence of driving, an aggressive display, and facilitates the appearance of sexual-aggressive components generally, while exogenous androgen facilitates a more natural sequence with bowing progressing to the more sexual manifestations of mounting and copulation. Hutchison's data (1970b) indicate that the chasing component of the display persisted in castrates, whereas bowing was eliminated, but the extent to which exogenous FSH will evoke displays in castrates has yet to be determined. It does seem probable that it may act synergistically with androgen. Moreover, on the basis of behavioural evidence, it is feasible that FSH is involved in the release of androgen. Exogenous LH had little effect on behavioural sequences in this species, but it did increase the probability that attacking would follow bowing. In *Quelea quelea*, although androgen injections would not increase the social status of low rank birds kept in small groups, exogenous LH would. Similarly, LH and not androgen seems to determine social status in *Sturnus vulgaris*.

Natural selection will favour males who can fertilize a potentially good mother, while females must ensure they have adequate protection from a mate and territory before accepting a male's advances. Initially the male's behaviour is aggressive and centers on depositing his sperm in the female before another male can do so; hence copulation occurs early in the behaviour sequence when aggressive behaviour is paramount. Females will usually squat for a male only in a territory in which they also have become accepted, but this does not prevent males making sexual advances to all potential mates when on the feeding grounds and away from a territory. So, selection tends to advance the time of copulation for the male, retard it for the female, and results in a compromise balance. Having fertilized a female, a male must now ensure adequate protection for the forthcoming eggs and young; and to a large degree, the remaining needs of his courtship sequence are, so to speak, to mollify his partner and provide her with suitable nesting facilities. But it appears that the male is initially endowed with high titers of plasma androgen (and probably also of FSH) which would be detrimental to the progression of the cycle. In fact, it appears that feedback mechanisms become initiated in consequence of courtship and lead to a reduction in androgen levels.

It is known that implanted testosterone propionate in the preoptic region of the hypothalamus reestablishes copulatory behaviour in capons so long as the necessary external stimuli are provided, and such work is beginning to define the hormone-sensitive regions of the central nervous system which must underlie sexual behaviour. This led Hutchison (1970b) to examine whether other sexual patterns of courtship preceding copulation might be linked to androgen-sensitive areas of the central nervous system, especially since there had been a suggestion that aggressive and copulatory activity might depend on separate androgen-dependent mechanisms in the central nervous system. Hutchison, therefore, investigated the effects of intracerebral implants of crystalline testosterone propionate on castrated Ring Doves; he was careful to segregate his experimental subjects according to their experience and preparedness to perform the nest-demonstration display as a sequel to the bowing display. The type of courtship evoked and whether fragmentary or the complete sequence of chasing, bowing, and nest soliciting (nest demonstration) depended on the anatomical position of the implant. Only those males with anterior hypothalamic and preoptic implants showed the complete reestablishment elbowing, while implants in the area basalis and hypothalamicus posterior medialis resulted in maximum nest soliciting. Hutchison's data (1970b) are consistent with the hypothesis that the stimulatory effects resulting from posterior hypothalamic implants depended on the diffusion in low concentration of testosterone to an androgen-sensitive area in the preoptic and anterior hypothalamic area; i.e., mechanisms underlying chasing and bowing require higher hypothalamic concentrations of androgen than mechanisms underlying nest soliciting. This evidence does not support the view that different androgen centers control different behaviours. Nonetheless, separate sites are possibly involved if a wider range of behaviour is considered. Electrical stimulation of the preoptic and anterior hypothalamic areas of the forebrain of pigeons induced an intensive bowing with all normal components, and this display could then become transformed into nest demonstrations via displacement preening. Defense and escape behaviour in the same species was induced by electrical stimulation of di- and telencephalic structures in the forebrain.

Following castration, bowing and chasing are the first displays to be eliminated while nest demonstration persists for longer, suggesting that it depends on a lower threshold of androgen stimulation. Indeed, Hutchison (1970b), using different sized hypothalamic testosterone implants, found that high-diffusion implants, delivering high concentrations, resulted in longer durations of bowing than smaller medium-diffusion implants, while low-diffusion implants induced nest soliciting in the virtual absence of chasing and complete absence of bowing. This at first suggests a simple and plausible explanation for the mechanism underlying the natural display sequence. But Murton et al. (1969b) also found that nest demonstration became fully established, albeit after a delay, in subjects maintained on high levels of exogenous testosterone treatment, so that these high dosages did not inhibit the onset of nest demonstration. Instead, these data suggest that a new mechanism becomes activated that can nullify the effects of exogenous androgen. The exogenous administration of estradiol monobenzoate immediately and consistently caused the appearance of nest-demonstration displays in intact feral pigeons and also in Ring Doves, while hypothalamic implants similarly induced long durations of nest demonstration in castrated doves. We do not know whether estrogen is involved in the normal cycle of the male, but the capacity to synthesize estrogenic steroids by avian testicular tissue is well established.

There is other evidence that androgen levels may decrease during the preincubation display phase of the feral pigeon. Lofts et al. (1973b)

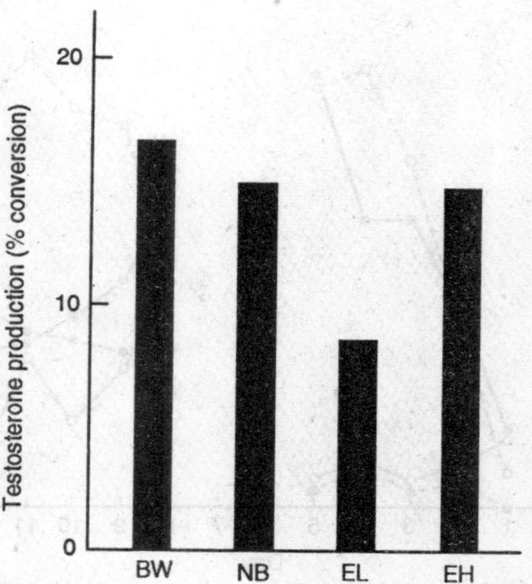

Fig. 4.7. The in vitro production of testosterone from radioactive pregnenolone by the pigeon testis during the bowing (BW), nest building (NB), egg laying (EL) and hatching (EH) periods of the reproductive behaviour cycle.

incubated testicular and ovarian material taken from feral pigeons at different stages of the reproductive cycle with tritiated pregnenolone as the steroid precursor. The synthesizing ability of the gonads at different stages was then measured using thin-layer chromatography to separate the different steroids produced. During the initial period of bowing display, there was a peak in androgen synthesis (i.e., during the first 3 days of the cycle), but approximately 10 days later and just prior to egg laying, androgen synthesis was markedly reduced. An increase in progesterone levels was also noted at this time. The capacity of the Leydig cell to synthesize a substance need not imply that the same substance is released from the cell, but it is almost certainly significant that quite separate evidence points to the existence of a high plasma titer of progesterone prior to egg laying in Ring Doves. Thus, males of this species that have previous breeding experience can be induced to incubate eggs by injections of progesterone or, in a smaller percentage of cases, following prolactin administration. All the same, when progesterone-injected and prolactin-injected doves were tested in bisexual pairs, more incubated than when tested alone, so the presence and behaviour of a mate significantly influences hormone-induced incubation responses.

Progesterone crystals chronically implanted into the brains of reproductively experienced Ring Doves induced incubation behaviour if placed in the preoptic nuclei and lateral forebrain, while the sexual

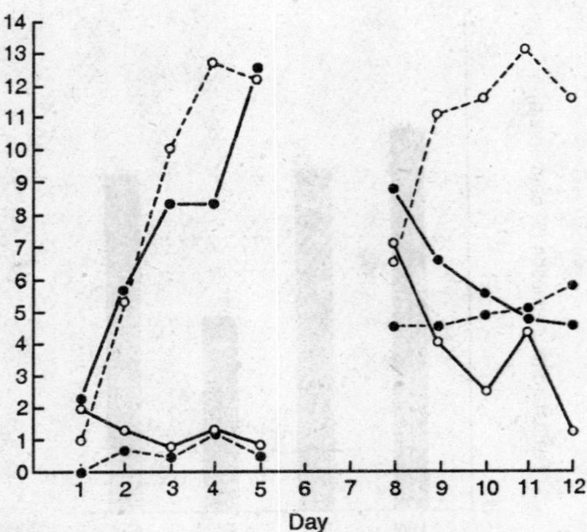

Fig. 4.8. Daily incidence of nest demonstration by paired male pigeon subjected to different hormone treatments.

and aggressive components of male courtship were suppressed. Male castrates, caused to exhibit courtship displays by treatment with testosterone, were selectively inhibited from bowing and induced to display nest soliciting alone when simultaneously injected with progesterone. So it seems conceivable that rising progesterone levels could inhibit androgen effects leading to the suppression of sexual-aggressive displays and instead allow the emergence of pair-cementing displays, such as mutual allopreening (caressing) with the onset of nest building and egg laying. But we must still not neglect the possibility that an estrogenic phase occurs in the male pigeon during the middle part of the preincubation cycle, and certainly exogenous estrogen is more effective than exogenous progesterone in causing nest demonstration.

A changed endocrine state during the preincubation cycle must depend either on exhaustion of a mechanism already engaged, which seems unlikely, or on the specific stimulation of neural-hormonal pathways elicited by specific behavioural events in the cycle. This second proposition is more likely in allowing for changes consequent to the natural or artificial disruption of the cycle. The preferential production of progesterone, for example, must depend initially on neural or hormone mechanisms initiated at higher levels of integration. We do not yet have the means to measure plasma FSH levels accurately and can only suspect that FSH titers decline during the preincubation behaviour phase. The spermatogenetic sequence in feral pigeons shows that numerous spermatozoa are free in the tubule lumen at the start of the cycle and there exists a peak in the number of spermatids. By the middle phase, the spermatozoa lying free in the tubule lumina have disappeared, whereupon the existing spermatids mature into a new generation of spermatozoa at about the time of egg laying. Thereafter, there is a new wave of spermatogenetic activity, with numerous divisions of primary spermatocytes leading to a new production of spermatids during late incubation. Plasma LH titers have also been measured using a radioimmunoassay technique, and they also show a fluctuating titer during the preincubation period. Whatever the final answer proves to be, it seems highly probable that a changed pattern of gonadotropin secretion might be elicited by specific behaviours or other external stimuli and that a variable and probably synergistic balance of gonadotropin output would then result in a differential steroid production by the gonads.

Craig (1911) was the first to demonstrate that the caressing display of the male pigeon could facilitate ovulation in the female. Matthews (1939) later showed, by separating males and females with a glass plate, that the stimulus causing ovulation was visual and not tactile. Subsequently, Erickson and Lehrman (1964) found that castrated males separated from females by a glass plate were far less effective in stimulating oviduct development in the female than were intact birds. The elegant work of Lehrman and his colleagues has now well established that stimuli arising from association with a mate cause the birds to become interested in nest building, and that once in this state the stimuli produced by the presence of nest material facilitate the birds becoming interested in sitting on eggs.

The behavioural acts involved in nest building by the male stimulate gonadotropin secretion (presumably FSH) in the female, causing gonadal hormone secretion from the ovary, which in turn leads to oviduct development. Thus, in both pigeons and canaries, nest-building activity is temporally associated with rapid oviduct development and ovulation followed a few days later by oviposition and the onset of incubation. In both doves and canaries, estrogen causes a moderate increase in oviduct weight, and whereas progesterone alone has no effect, it potentiates the effects of estrogen to produce maximum oviduct development. Prolactin can also augment the estrogen response, possibly by causing progesterone release. But while exogenous estrogens readily cause oviduct development, very high doses are necessary to initiate actual nest building in doves and canaries. Paired and photostimulated canaries in winter show nest-building behaviour, while birds injected with PMS (pregnant mare serum) can lay without nest building. These observations support the likely situation that gonadotropin secretions potentiate nest-building behaviour, which exogenous progesterone or prolactin certainly do not.

In the female canary, incubation-patch development occurs coincidentally with nest-building and oviduct development. Lehrman's studies make it clear for doves that exogenous prolactin, unlike progesterone, will not initiate incubation behaviour but that prolactin can maintain such behaviour once initiated. Moreover, judging from the development of the crop gland in pigeons, prolactin secretion is stimulated by the specific act of incubating eggs, though prolactin secretion can be induced in the male if he can see the incubating female and if he has previously associated freely with her during earlier courtship. The appearance (artificially) of squabs once the female has laid eggs can also accelerate crop growth and, presumably, prolactin secretion. The pituitary glands of domestic hens contain prolactin only so long as they are allowed to sit on eggs.

To summarize, the preincubation reproductive cycle of the female appears to involve a period of increasing estrogen production, which it is tempting to attribute to increases in FSH production. Next follows a phase during which progesterone or prolactin secretion, or both, become involved, and the important endocrine problem needing to be resolved is whether increasing estrogen levels give rise to progesterone production which in turn stimulates prolactin secretion, or whether prolactin secretion is first stimulated leading in turn to progesterone production. Since Meites and Turner (1947) found no increase in pituitary prolactin level following progesterone injection, and since Lehrman's studies (1963) reveal no crop sac development in pigeons following progesterone administration, we may prefer the second alternative or acknowledge the possibility that nonmeasurable levels of prolactin were involved.

In the male feral pigeon, administration of exogenous estrogen causes (1) a regression of the testis tubules with only spermatogonia remaining present; (2) the encapsulation of the old Leydig cell generation, which becomes densely lipoidal; and (3) the appearance of a new Leydig cell generation, also heavily lipoidal. Coincidentally, plasma LH titers are markedly elevated, and it seems likely that the histological manifestations depend on the known elevation of LH coincident with a presumed suppression of FSH activity. In the male feral pigeon, LH titers increase during the end phase of the prelaying cycle and after the emergence of nest-soliciting displays. While it has yet to be resolved whether estrogen is liberated during the male's prelaying cycle (though it is known that the testis can produce estrogen), it does seem reasonable to anticipate that rising estrogen production in the female suppresses FSH activity and stimulates LH production (the Holweg effect) and that such a change in gonadotropin balance provides the necessary conditions for progesterone (prolactin) secretion. It is of course known that there is a boost of LH production associated with ovulation in the domestic hen. Analogous endocrine changes apparently occur in both male and female, particularly at the hypothalamic level, though the male cycle is advanced compared with that of the female. In the male, an initial FSH androgenic phase gives way to an LH and progesterone phase leading to incubation. In the female an FSH-estrogen phase is stimulated in consequence of the males' courtship, and this leads to oviduct development relatively late in the cycle; it is probably significant that in feral pigeons the female assumes the more aggressive and dominant role coincident with the nest-building phase, and she may even drive the male away at this stage of the cycle. Significant developments in this field can be expected in the next few years.

5

HORMONES REGULATING OVULATION

Ovulatory cycles of birds in general, and of domestic chickens in particular, are not comparable to the reproductive cycles of vertebrates which are either above or below birds on the evolutionary scale. In most polytocous vertebrates, groups of follicles mature together, reach ovulatory size together, and are destined to ovulate together. Episodes of simultaneous multiple ovulations are succeeded by periods of growth, maturation, and ovulation of new crops of follicles until pregnancy or the end of the breeding season intervene. Thus, most polytocous vertebrates other than birds are cyclically polytocous. Birds, on the other hand, can be looked upon as being continuously polytocous because in them growth and ovulation of follicles is not periodic, as it is in mammals, but continuous within breeding seasons. In mammals cyclic ovarian function has been explained reasonably well, by invoking endocrine feedback mechanisms through which hypophysis and ovaries regulate each other. The endocrine problem confronting birds is basically different from the one faced by mammals. In chickens, endocrine feedback mechanisms, like those existing in mammals, probably play a secondary role and may even be completely absent. In mammals, however, there are mechanisms which must *assure* the periodic and simultaneous maturation and ovulation of many eggs; in chickens there are mechanisms which must *prevent* the maturation and ovulation of more than one egg at one time. The problem in birds becomes essentially one of providing an endocrine mechanism permissive of the existence of a hierarchy of follicles of graded sizes, only one of which is capable of ovulating at any one time. To build such a hierarchy of follicles of graded sizes by injection of hormones into birds turns out to be a difficult problem for the investigator, although it is normally easily accomplished by laying birds. Injection of graded doses of gonadotrophic hormone into intact laying hens which possess a complement of small follicles of assorted sizes shows that exogenous hormones can hasten the maturation of large numbers of follicles. If the level of injected hormone is high, ovulatory size will be reached by several follicles simultaneously, whereas at lower doses follicles are caused to increase in size until a great many of them are within the same size and weight range, even though none of them may reach ovulatory size. Thus, the ultimate number and size of follicles stimulated to precocious growth by exogenous gonadotrophins depend on the dose of hormone injected. It is noteworthy that, as a rule, size gradation between follicles will be obliterated or at least substantially reduced regardless of whether follicles are sufficiently stimulated to grow to ovulatory size. These experiments show that the smaller follicles in the graded series are perfectly capable of responding to hormonal stimulation by growth and eventually by ovulation, and that the degree of stimulation obtained is proportional to the amount of exogenous hormone injected.

How size gradation is maintained in normal untreated birds is not known, although the following theory appears plausible. In examining the circulatory system of the chicken ovary, one is impressed by the extreme complexity of that system and by the fact that the vascular networks covering the large follicles are disproportionately more extensive and intricate than the networks supplying the smaller follicles. Ovulation of the largest follicle of the series results in the abrupt shutting down of a very extensive circulatory network supplying this follicle. The amount of blood that can flow through the networks of other, smaller follicles increases after ovulation. If it is assumed that the pituitary gland of chickens releases a steady and constant amount of gonadotrophic hormones, it seems probable that this constant amount of hormone is divided between follicles in accordance with the degree of development of their respective vascular systems. Thus, the degree of hormonal stimulation available to each follicle would depend on the extensiveness of its vascular system. After the largest follicle ovulates, the available hormone is distributed among the remaining follicles, each of them getting quantities of hormone according to its capacity as determined by the extensiveness of its circulatory system. How, under this scheme, some follicles get a head start on their mates is not quite clear, although this could be a simple matter of chance and could depend completely on the initial proximity of some follicles to crucial blood vessels.

Although portions of the preceding discussion are based on conjecture, experimental evidence available at present supports the assumption that the pituitary gland of chickens usually pours out a steady and unvarying stream of gonadotrophic complex consisting of FSH- and LH-like substances which are distributed among the follicles in the manner suggested above. The evidence for a steady flow of gonadotrophic hormone comes in part from the following experiments.

We have already seen that gradations in follicular size are abolished or lessened it exogenous gonadotrophins are injected. This observation suggests that the rate of flow of endogenous gonadotrophins (or of their utilization), is normally adjusted in such a way as to permit the maturation of a graded series of follicles. The addition of exogenous hormone increases the total amount of circulating hormones to the point where even the smaller follicles, which normally would not get enough stimulus to cause rapid growth, can grow much more rapidly. Experimentally it appears difficult to gauge the quantity or quality of exogenous hormone in such a way as to speed up the growth rate of individual follicles without abolishing the normal gradation in size. These experiments, however, do suggest that the factor limiting the growth of follicles in number and size is the amount of gonadotrophic hormone available to each follicle.

A second series of experiments also supports the theory of a constant flow of gonadotrophic hormones. If, in laying hens, a foreign body such as a bead or even just a thread is inserted into, or passed through the lumen of the albumen-secreting part of the oviduct, ovulations are stopped almost completely in the majority of hens. A similar effect is obtained if a thread is inserted into the isthmus of laying hens, although the uterus does not appear to participate in the neurogenic inhibition of LH release. The interest in these experiments is centered on two facts. The first is the cessation of ovulations, an effect to which we shall return later. The second is of immediate interest in that it is found that in spite of cessation of ovulations for as long as three weeks, neither the ovaries, oviducts, or combs regressed or even diminished in size during this period. The ovarian follicles during this anovulatory period were maintained at the sizes which they had reached at the time of insertion of the foreign body into the oviduct. No follicular atresia was evident during the first three weeks of the anovulatory period.

Because during this time neither the oviduct nor the comb decreased in weight, we are forced to conclude that the ovary continues to secrete amounts of estrogen and androgen adequate for the support of comb and oviduct. Because the only effect of the presence of the thread in the oviduct is the absence of ovulations, the conclusion is drawn that we are dealing with a neurogenic inhibition of the release of ovulatory quantities of the ovulation-inducing hormone from the hypophysis. This theory was tested by injecting these hens intravenously with LH of mammalian origin. When this was done, it led to the ovulation of a single ovum. The assumption that the presence of a foreign body in the oviduct of a hen selectively inhibits the release of LH alone from the animal's own pituitary gland thus appears justified.

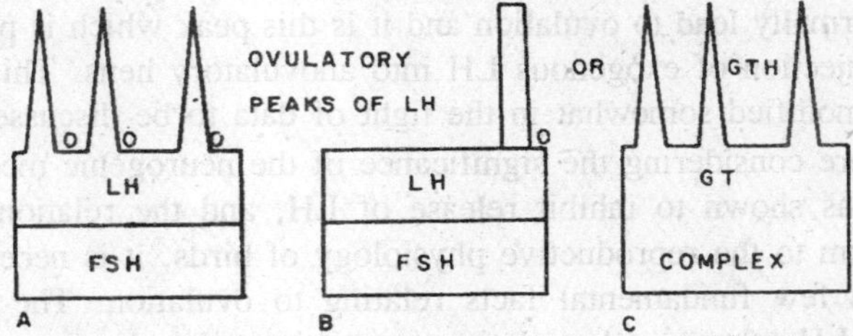

Fig. 5.1. A model of hypophyseal function in hens suggested by experiments involving the presence of foreign bodies in magnum.

Next we note that hypophysectomy or sectioning of the hypophyseal stalk in laying hens results in immediate reduction in ovarian size, follicular atresia, and regression in comb size and oviduct weight. Since none of these effects followed the insertion of a thread into the oviduct of laying hens, this demonstrates that there was no interruption or diminution in rate or amount of flow of gonadotrophins essential for support of these functions. Furthermore, if the largest follicle of hens with oviducal threads is forced to ovulate by injection of LH, the whole hierarchy of follicles of less than ovulatory size moves up one size notch and fills the size vacancies created by the ovulation. That this would happen in the absence of continuous hormonal stimulation is highly improbable.

This, in brief, is the basis for the theory that the pituitary gland of birds secretes a steady and unvarying amount of gonadotrophic complex which, when tested on mammals, appears to have the properties usually ascribed to a mixture of FSH and LH.

It should be noted that in hens with oviducal threads growth of smaller follicles is suspended during the anovulatory period and that it is resumed only if the largest follicle has been caused to ovulate. Under the constant and steady flow theory one might expect the smaller follicles to continue to grow and eventually catch up with the largest one, thus obliterating the normally observed size gradation. That this does not happen can be explained by invoking the previous assumption that the total amount of stimulus to which each follicle is exposed is limited by its blood supply. Only after the largest follicle is shed does the amount of hormone previously used by it become available to the smaller members of the group.

Finally, in order to account for the occurrence of ovulations in normal laying hens, it can be assumed that at certain intervals the pituitary gland releases quantities of gonadotrophic complex over and above the base amount which is released continuously. These ovulatory peaks normally lead to ovulation and it is this peak which is provoked by the injection of exogenous LH into anovulatory hens. This theory will be modified somewhat in the light of data to be discussed next.

Before considering the significance of the neurogenic mechanism which was shown to inhibit release of LH, and the relation of this mechanism to the reproductive physiology of birds, it is necessary to review a few fundamental facts relating to ovulation. The interval between LH release and ovulation was estimated to be from 8 to 14 hours by Van Tienhoven et al. (1954). This estimate was obtained by using Dibenamine (*N,N*-dibenzyl-β-chloroethylamine) which is able to block both normal and progesterone-induced ovulations most completely when injected 8, 10, and 14 hours before expected ovulation. Fraps and his co-workers estimated this interval to be only 8 hours or less, basing their results on the ability of LH to induce ovulation in hypophysectomized hens. Which of these estimates is correct is of no great importance for the present discussion. For purposes of this discussion and for reasons of simplicity we accept the Van Tienhoven estimate, according to which 10 to 14 hours elapse between LH release and ovulation. Van Tienhoven (1955) further found that LH release in hens lasts from 26 minutes to $2^1/_2$ hours and that stimulation of the hypophysis and LH release take place concurrently. For comparison it should be noted that rabbits require about one hour, and rats one-half hour to release enough LH to cause maximal ovulation response. In these mammals, as in birds, hypophyseal stimulation and LH release are concurrent.

In order to understand better what is to follow, a few explanatory statements concerning the reproductive peculiarities of chickens are necessary. The egg spends 24 to 30 hours in the oviduct and only 4 to 5 hours of that time is spent in the magnum and the isthmus, the regions which were found to be sensitive to the presence of foreign bodies. The remainder of that time, the egg spends in the uterus, acquiring a hard shell.

A series of eggs laid daily and consecutively without an intervening rest period is called a clutch. Clutch performances of three typical chickens are given below, the numbers referring to the number of hours elapsing between ovipositions, the dashes indicating days on which no eggs were laid.

1. —49—45, 28—45—47—47, 29—43, 28
2. —43, 27, 27, 29—43, 29—50, 30, 27, 28—29, 24, 29, 29
3. 28, 25, 23, 25, 24, 25, 23, 25, 28, 24, 24, 25, 24, 30—45, 28

The performance of hen No. 2 is typical of the majority of laying hens, but neither the short clutches nor the clutches shown by hens Nos. 1 and 3 are unusual. Under ordinary lighting conditions ovulations and ovipositions always take place during daylight hours. Because

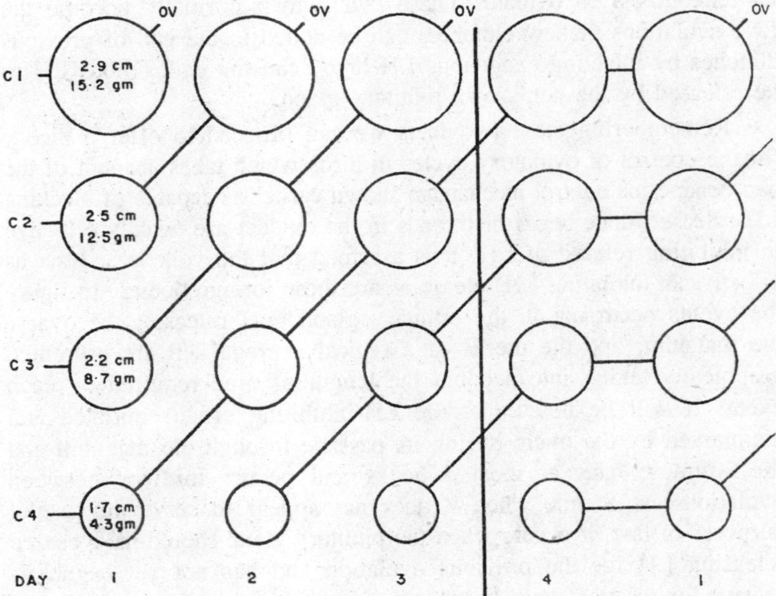

Fig. 5.2. Pictorial presentation of a clutch which ends at vertical line.

ovulations usually occur shorty after oviposition, and because complete egg formation usually requires more than 24 hours, hens like No. 2 ovulate each succeeding egg later than the previous one, the last egg of the clutch being ready for laying by 3 P.M. or, at the latest 4 P.M. of the last day. When this happens the ovum that is destined to ovulate next is not shed, and, in fact, its ovulation is delayed. It will ovulate early in the morning some 40 to 50 hours after the last previous ovulation and this ovulation initiates the next clutch.

Two facts are important in this context. First, it seems as if light were the deciding factor in determining the time of ovulation. This conclusion, however, is not justified because in hens exposed to continuous light, ovulations occur evenly spaced throughout the whole 24-hour period of light. In the regular alternating sequences of 12 hours of light and 12 hours of dark, LH is released predominantly during the hours of darkness, but, when hens are exposed to continuous light, LH continues to be released without regard to the times of LH release adhered to during the preceding light-dark sequences.

The second fact to keep in mind is that failure of ovulation of the next egg in the hierarchy at the end of a clutch sequence is not due to its unreadiness to ovulate. These ova (which normally become the first ovulations of new clutches) can be added to the end of previous clutches by injecting exogenous LH or by causing endogenous LH to be released by the hen's own pituitary gland.

Remembering these two facts we can proceed to offer a theory for the control of ovulatory cycles in birds which takes account of the neuroendocrine control mechanism shown earlier as capable of blocking LH release. Since beads or threads in the oviduct are equally effective in inhibiting release of LH, it is assumed that the yolk is at least as effective in inhibiting LH release as are these foreign bodies. In figure the events occurring in the pituitary gland (LH release), the ovary, the magnum, and the uterus of a typical average hen are presented graphically taking into account the length of time required by each event. It will be noted that the LH-inhibiting effect, initiated and maintained by the ovum during its passage through the magnum and the isthmus, lasts at most 5 hours and occurs midway between ovulations, at a time when it does not appear to serve any useful purpose. In fact, it occurs when the pituitary gland should have ceased releasing LH for the previous ovulation and has not yet begun its release for the next one. It appears improbable that this phenomenon is an artifact which plays no role in the reproductive physiology of birds. It is postulated, therefore, that at the beginning of a clutch a

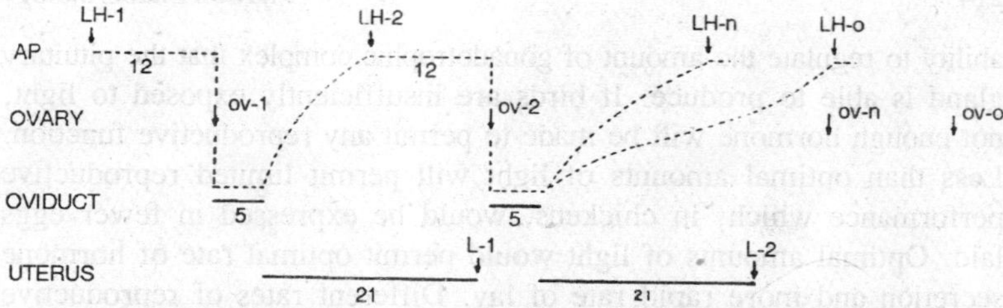

Fig. 5.3. Time intervals elasping between LH release, ovulation, egg-formation and laying.

chicken secretes and continuously releases LH at a level at which the hormone is able to induce ovulation of the largest follicle. Immediately after ovulation the ovum is engulfed by the oviduct and initiates the neurally mediated depression of LH secretion and release. This control is maintained for about 5 hours. When the egg passes from the isthmus into the uterus, the brake on LH secretion is released. The rate of secretion and release of LH slowly returns to pre-inhibition levels. When the pre-inhibition level of LH is reached, the next egg of the clutch is ovulated. With each succeeding cycle of inhibition and release hypophyseal recovery rate becomes slower until at the end of the clutch the pituitary gland does not recover in time to cause the ovulation of the next egg, and the clutch is interrupted. In this scheme, clutch length would be determined by the rate at which the pituitary gland could recover from each episode of neural inhibition. Thus, rapid recovery would permit long clutches, while slow recovery would result in short clutches.

Fraps (1946) has reported that the first follicle of a clutch is significantly more sensitive to exogenous LH than are the follicles destined to ovulate later. Bastian and Zarrow (1955) have confirmed and extended this finding. An alternate explanation, rests on the assumption that the recovery rate of the pituitary gland is the same after each cycle of inhibition; the daily delay in time of ovulation of successive clutch mates is caused by their progressively increasing insensitivity to LH. Increasing follicular insensitivity to LH makes it difficult to explain why many hens can have extremely long clutches consisting of up to 366 eggs. This difficulty is obviated if clutch length is determined by the recovery rate of the pituitary gland. It is, of course, possible that both the rate of recovery and follicular sensitivity play a role in determining clutch length.

On the basis of available data it seems improbable that light, as such, plays a role in governing the rhythmicity of the ovulatory cycle in birds. It appears most probable that the main function of light is its ability to regulate the amount of gonadotrophic complex that the pituitary gland is able to produce. If birds are insufficiently exposed to light, not enough hormone will be made to permit any reproductive function. Less than optimal amounts of light will permit limited reproductive performance which, in chickens, would be expressed in fewer eggs laid. Optimal amounts of light would permit optimal rate of hormone secretion and more rapid rate of lay. Different rates of reproductive performance in a group of birds exposed to identical light conditions may be attributable to a genetic difference in light sensitivity among members of the population. Experimental evidence supporting these contentions is now being accumulated.

Although light determines the rate of hypophyseal function and thus has a "permissive effect" on reproductive performance, the rhythmicity of the laying cycle of birds does not appear regulated by light but by neurogenic inhibition and release of the gonadotrophic complex. This cycle of alternating neurogenic inhibition and release of hormone secretion functions equally well under a variety of light regimes. This is true, however, only within those limits of illumination which assure that the pituitary will be stimulated by "some" light, so that it can secrete "some" gonadotrophic hormone.

Recent studies by Ralph and Fraps (1958) strongly suggest that a neurohumoral relation exists between hypothalamus and pituitary gland. Using the well-known ability of progesterone to induce LH- release and hence ovulation, in hens, these authors show that electrolytic lesions placed in the anteroventral hypothalamus usually prevent progesterone-induced ovulations which, according to them, occur 7 or 8 hours after progesterone injection. Ovulations are prevented if the lesions are made within about 2 hours after progesterone injection, but ovulations follow if the interval between progesterone injection and placing of lesion is greater than 2 hours. Thus, progesterone-induced release of LH is mediated through a neural pathway and it appears that the neural mechanism participates in LH release for about 2 hours.

6

Reproductive Adaptation

Under original primitive conditions, particularly in wild birds, reproduction passes through sequences of distinct phases, alternating mainly between a period of sexual rest (eclipse) and one of sexual activity (nuptial or breeding condition). In many species mating and egg laying are restricted to a single period of not more than one week per year, whereas other activities, such as courtship, nest building, incubating, and taking care of offspring are consuming the larger part of the breeding season.

These seasonal reproductive cycles offer an abundance of spectacular examples of correlative adaptation, the meaning or evolutionary value of which remains often entirely obscure. The pivotal phenomenon is the yearly recrudescence and subsequent regression of the sex glands during the breeding season. It involves 200- to 500-fold weight increases with concurrent histologic maturation of testes and ovaries during the month preceding actual reproduction. These changes are equivalent to a progress from the juvenal to the mature condition. Regression with degeneration of not discharged fully or partially matured gametes begins even during incubation, and nurturing of nestlings, and continues until the inactive or eclipse condition is again established.

Associated with gonadal sex and sex cycles are a great variety of other manifestations, some of which *control gonadal development*, while others are *directed by gonadal hormones*. However, it becomes also increasingly evident that certain periodic changes, though being part of the general sex complex, take their course *independently of gonadal hormones*, even in castrates.

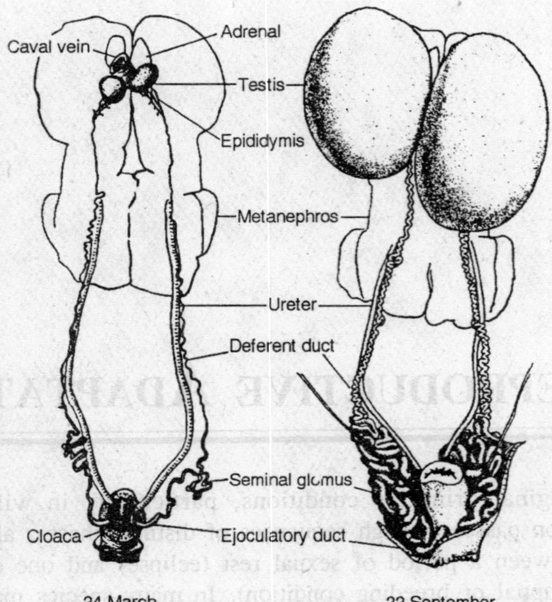

Fig. 6.1. Euplectes f. franciscanus, orange weaver finsh. Urogenital organs of males in eclipse and in nuptial condition.

Control of Gonadal Cycles

Gonadal growth and ovulation are directly controlled by hypophyseal gonadotrophic hormones. Like the gonads of nestling birds, the small testes and ovaries of the adult in early eclipse season are quite refractory to stimulation by injected hormones. But with the approach of the breeding season, usually beginning in October or November, they become responsive and enlarge in proportion to the amount of the injected gonadotrophin.

The problem of the *synchronization* of the gonadal cycle with the seasons of the year has of course long been recognized. But only since Rowan (1929) showed that an artificial increase of the daily lighting period in winter causes the small eclipse testes of song birds to recrudesce precociously have experimental endocrinologists started to investigate the nature of these reproductive rhythms. This work culminated in the discovery that hypothalamic centers by means of neurohumoral mechanisms exert some control over at least some of the gonadotrophic functions of the hypophysis. The subject has recently been reviewed by Marshall (1955) and Benoit (1955). In general it is

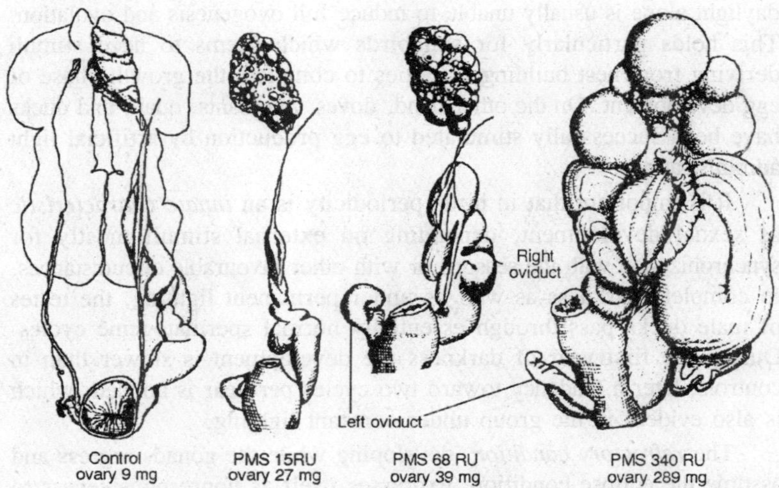

Fig. 6.2. Passer d. domesticus L., domestic sparrow. Sex organs of females.

true that exposure to extended lighting has a stimulating effect, which eventually results in increased gonadal activity of birds as well as other vertebrates. However, day-length is only one of many external factors with which the year cycle may become associated. It plays a minor role, if any, in equatorial birds. Even in relatively stationary inhabitants of northern regions, such as the domestic sparrow, increased

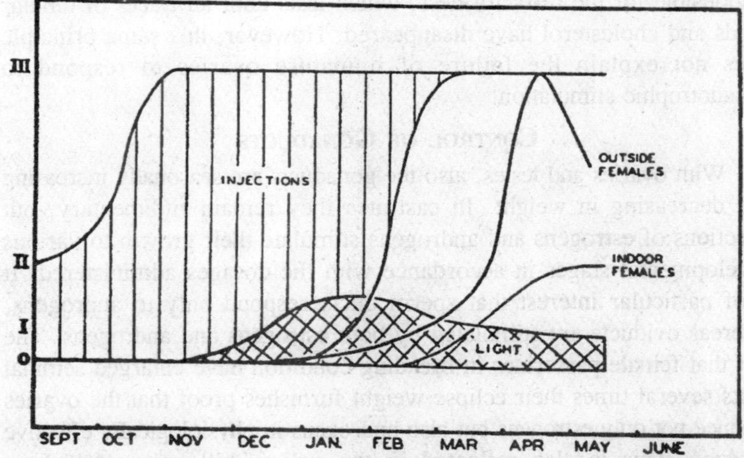

Fig. 6.3. Development of ovaries in the female sparrow under normal and experimental conditions. O, eclipse condition; III, full breeding condition; I and II, intermediate developments.

daylight alone is usually unable to induce full ovogenesis and ovulation. This holds particularly for songbirds which seems to need stimuli deriving from nest building activities to complete the growth phase of egg development. On the other hand, doves, pheasants, quail, and ducks have been successfully stimulated to egg production by artificial light administrations.

It is important that in birds periodicity is an *innate characteristic* of sexual development, depending on external stimuli mostly for synchronization with the seasons or with other favourable circumstances. In complete darkness as well as under permanent lighting, the testes of male ducks pass through essentially normal spermatogenic cycles. During the first year of darkness the development is slower than in controls; later a tendency toward two cycles per year is noticed, which is also evident in the group under constant lighting.

The *refractory condition*, developing while the gonads regress and assume the eclipse condition, expresses itself as nonresponsiveness to injected gonadotrophins and to light stimulation. The French authors assume that it is the consequence of a "fatigue" of the gonad stimulating mechanism which develops after a period of activity. Unfortunately they do not specify the level at which this fatigue develops: gonad, hypophysis, or hypothalamus. Marshall (1949, 1955) has accumulated much evidence favouring the view that fatty metamorphosis within the postnuptial gonads (regressing seminal tubules, atretic follicles) is responsible for the refractoriness, which lasts until all traces of tubular lipids and cholesterol have disappeared. However, this same principle does not explain the failure of immature ovaries to respond to gonadotrophic stimulation.

Control of Gonaducts

With ovaries and testes, also the gonaducts are seasonally increasing and decreasing in weight. In castrates they remain rudimentary, but injections of estrogens and androgens stimulate their growth to various developmental stages in accordance with the dosages administered. It is of particular interest that sperm ducts respond only to androgens, whereas oviducts are stimulated by both estrogens and androgens. The fact that female passerines in breeding condition have enlarged seminal ducts several times their eclipse weight furnishes proof that the ovaries produce not only estrogens but also androgens in physiologically effective dosages. This is also reflected in the yellow bill colour which is assumed by female as well as male starlings during the breeding season, a pigmentary change that is controlled by androgenic hormones.

Control of Nuptial Plumages

The distribution of hen and cock plumages in the many species of birds is a very complex subject, and the mechanisms that control juvenal and adult plumage types have been studied in only a few instances. In connection with the problem of reproductive adaptations the situation of some African weaver finches, particularly those of the genus *Euplectes*, is of particular interest because it furnishes evidence of cycles of hypophysial activity in male and female castrates. After complete orchectomy the male continues year after year to change nuptial and eclipse plumages synchronously with intact males. The females, which normally wear always the eclipse, or hen, plumage, after complete ovariectomy also start to alternate seasonally between hen and cock plumages. Inasmuch as the nuptial plumage type is determined by the luteinizing hormone (LH), the autonomous cycle is only established for this one member of the group of gonadotrophins. Significantly it is the one that also induces ovulation, and its release is often regulated by hypothalamic centers. Therefore, it appears probable that the hypophysial cycle which directs the plumage changes is itself also under the control of some centers in the central nervous system.

Riddle of Migration

It is not at all surprising that the flights of birds to and from their breeding places have caused many observers to speculate about a possible hormonal linkage with the sex cycle. However, as pointed out in a recent review by Farner (1955), serious difficulties and contradictions appear whenever such theories are submitted to critical tests. Not only would it be unavoidable to assume that spring and fall migrations are determined by opposite hormonal states, but reports on migration of castrated birds constantly increasing in number prove to the contrary that hormonal support is unnecessary. Looking at the situation without preconception, it actually does not seem to call for any hormonal mediation at all. The general pattern of migration, changing from species to species, is inherited and only the immediate release of the urge to travel is in question. Although certain conditions of physical fitness must enter as permissive factors, it is quite obvious that environmental changes supply the immediate cause of the decision to change living spaces. No matter whether the external event, serving as the cue be a change in daylight, temperature, humidity, food supply, or any combination of such factors, engendered stimuli enter the brain by sensory pathways. Their impressions become associated with the latent behaviour patterns. This might just as likely be accomplished directly, by neurophysiologic processes, as by intervention of endocrine systems. Actually, the fact that hormone links have not been found, and that to all known appearances they do not play any role as basic control factors of migration, favours the first alternative. This does not minimize the importance of the coordination of migration with seasonal development of sexual characters in many species. However, the relationship, where existent, seems to be one of parallel development rather than of cause and effect.

7

HORMONES IN MIGRATION

Migration in birds is a complex behaviour pattern. It occurs periodically with remarkable temporal precision and comprises a sustained orientated movement over great distances from a breeding to a non-breeding area. Its temporal precision and its orientation pose the intriguing problem of regulation, and the studies of the past 35 years have demonstrated clearly the regulatory role of internal, or physiologic, factors and of external or environmental, factors. An analysis of bird migration is in reality a study of the ecologic and physiologic factors which initiate, maintain, and inhibit a complex behaviour pattern. The behaviour is thought to be adaptive, since it results in the seasonal use of favourable habitats in the higher latitudes for feeding and breeding.

Experimental studies of the regulation of migratory behaviour were initiated by Rowan (1925, 1929), who discovered that gonadal recrudescence could be induced out of season, in late fall and winter, by subjecting slate-coloured juncos (*Junco hyemalis*) to artificial increases in day length. On the assumption that migratory behaviour in the spring, when birds are flying north to their breeding grounds, was a phase of sexual behaviour. Rowan tested the effect of experimentally induced gonadal recrudescence on migratory behaviour by releasing juncos in winter, many months ahead of the normal time of their spring migration and with their gonads at various stages of development. From the results of these experiments. Rowan concluded that in the slate-coloured junco, and other fringillids, the stimulus to migrate in the spring was regulated by external and internal factors. The external factor was the increasing day length after December 21. The internal factor was the production of sex hormones, which he correlated with the recrudescing gonads of spring and the regressing gonads of fall. The minimal gonad of winter and the maximal gonad of the breeding period showed little or no hormone-producing interstitial tissue and, hence, would not stimulate migratory behaviour. From the results of later work with crows (*Corvus brachyrhynchos*), Rowan (1932) concluded that the southward migration in fall appeared to be independent of the influence of the gonads.

Rowan's epochal work defined the basic problems and stimulated an experimental attack on the regulation of bird migration and gonadal cycles. The studies performed since Rowan's discovery have centered around the following basic problems: (1) the nature of the external factor, (2) the physiologic changes induced by the external factor, and (3) the mechanisms whereby the external factor induces the physiologic changes which precede spring migration. The problems of how the change in physiologic state induces migratory behaviour and what terminates the behaviour have yet received little attention.

The first extensive experiments designed to test Rowan's hypothesis utilized the Oregon junco (*Junco oreganus*), a species closely related to the slate-coloured junco. These experiments corroborated Rowan's observation that birds could be stimulated to migrate northward months ahead of time by subjecting them to artificial increases in day length in the late fall and winter. However, they also demonstrated that birds already in breeding condition would migrate, a finding contrary to Rowan's conclusion. In addition to gonadal growth preceding migration, there was also a marked increase in body weight caused by large deposits of subcutaneous and intraperitoneal fat, which appeared to be a better criterion of a readiness to migrate than the condition of the gonad. Later studies demonstrated that the pituitary was also involved in this premigratory change in physiologic state. Members of a non-migratory race of the same species, which were exposed to the same environmental conditions in nature and in the laboratory, differed from migratory individuals in not showing marked deposition of fat nor increase in body weight and in having a much faster rate of gonadal development which resulted in an earlier breeding season.

The experiments of Rowan and Wolfson pointed clearly to the increasing day lengths of winter and spring as the environmental stimulus for spring migration, but there was a serious weakness in this aspect of Rowan's theory. Birds that wintered in the north temperate zone or northern subtropical zone would experience substantial increases in day length after December 21, but not the birds which wintered in

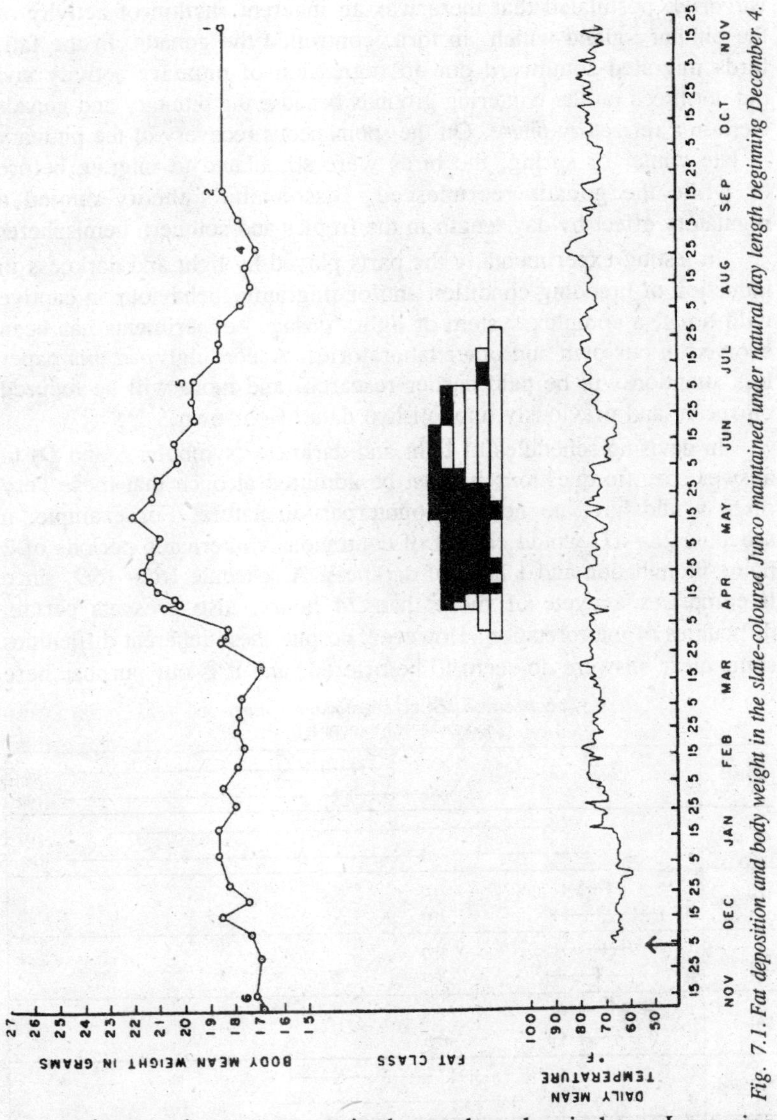

Fig. 7.1. Fat deposition and body weight in the slate-coloured Junco maintained under natural day length beginning December 4.

the tropics, on the equator, or in the southern hemisphere. Increasing day length, obviously, was not a feature of the environment on the wintering grounds of all migratory birds.

To overcome this weakness of Rowan's theory, Bissonnette (1937), on the basis of his studies of the gonadal cycle of the starling (*Sturnus vulgaris*), postulated that there was an inherent rhythm of activity of the pituitary gland which, in turn, controlled the gonads. In the fall, birds migrated southward due to regression of pituitary activity and did not breed on the wintering grounds because the pituitary and gonads were in a refractory phase. On the spontaneous recovery of the pituitary in late winter or spring, the birds were stimulated to migrate before or while the gonads recrudesced. Bissonnette's theory denied a regulatory effect by day length in the tropics and southern hemisphere.

In testing experimentally the parts played by light and darkness in induction of breeding condition and/or migratory behaviour in captive wild birds, a complex system of light "dosage" experiments has been evolved in our own and other laboratories. Accordingly, in this paper less attention will be paid earlier research, and more will be focused on recent and previously unpublished data of our own.

In devising schedules of light and darkness (symbols: *L* and *D*) to answer specific questions it must be admitted at once that these very often would have no normal counterpart in nature. For example, a schedule 2*L*—1*D* would consist of continuously alternated periods of 2 hours illumination and 1 hour of darkness. A schedule 16*L*—16*D*, since it comprises a cycle of more than 24 hours, also presents certain difficulties of interpretation. However, despite these inherent difficulties some clear answers do seem to be offered, and it is our purpose here

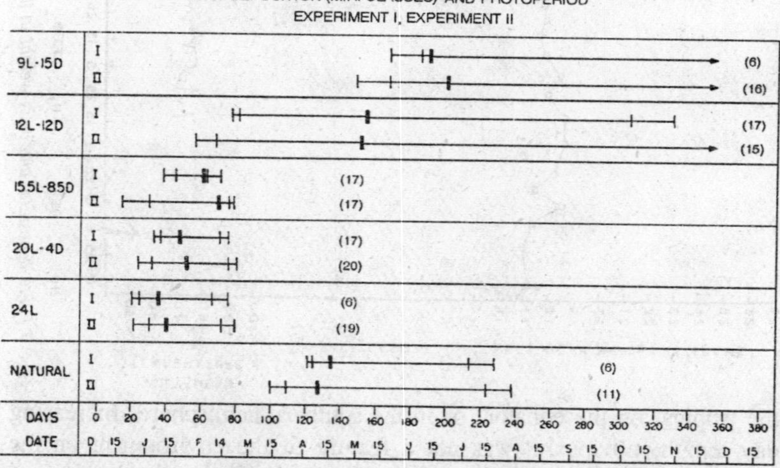

Fig. 7.2. Time of fat deposition in relation to photoperiod, beginning December 4. The upper horizontal bar in each photoperiod represents experiment I, the lower one experiment II.

to bring these out. As will emerge from the ensuing discussion, a lively debate centers about the relative contributions of alternated light and dark to gonadal maturation. With present criteria it has been impossible, obviously, to separate the contributions of each of these two physical states when one must always be tested in the presence of the other. In interpreting the results of "light dosage" experiments it is useful to bear in mind the possible mechanisms through which the experimentally altered physical conditions exercise their ultimate effects upon breeding condition. As is indicated by work such as that of Benoit and Assenmacher (1955) the nervous receptor in this case is the eye, and the nervous pathway probably leads through the hypothalamus in some way to the anterior pituitary. The eventual gonadal changes then follow gonadotropic hormone release.

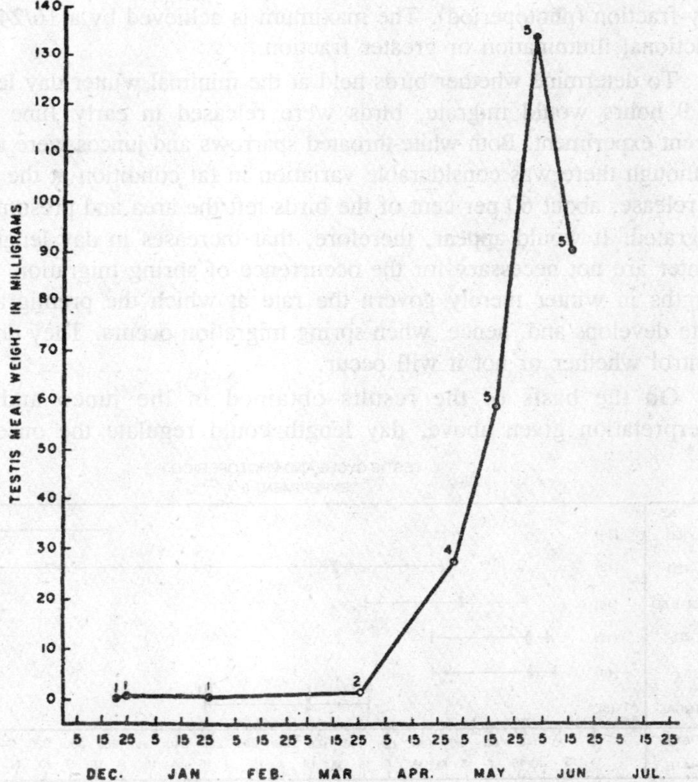

Fig. 7.3. Weight and development of testis in the slate-coloured junco under natural day lengths beginning December 4.

The species investigated most extensively in our laboratory are the slate-coloured junco and the white-throated sparrow, *Zonotrichia albicollis*. The criteria we have found most useful in judging positive responsiveness in a given experiment are body weight, size of depot-fat deposits, and gonadal size and gametogenetic stage.

Day Length and Initiation of the Migratory and Reproductive Responses

In agreement with others we were able to show with our birds that fat deposition and gonadal growth and maturation may be precociously stimulated by unseasonally increased illumination. The significant conclusions drawn from these experiments were that 9 hours per day or more of light were stimulatory, and that the rate and degree of stimulation were proportional to the length of the illuminated day-fraction (photoperiod). The maximum is achieved by a 16/24 day fractional illumination or greater fraction.

To determine whether birds held at the minimal winter day length of 9 hours would migrate, birds were released in early June in a recent experiment. Both white-throated sparrows and juncos were used. Although there was considerable variation in fat condition at the time of release, about 60 per cent of the birds left the area and presumably migrated. It would appear, therefore, that increases in day length in winter are not necessary for the occurrence of spring migration. Day lengths in winter merely govern the rate at which the premigratory state develops and, hence, when spring migration occurs. They do not control whether or not it will occur.

On the basis of the results obtained in the junco and the interpretation given above, day length could regulate the onset of

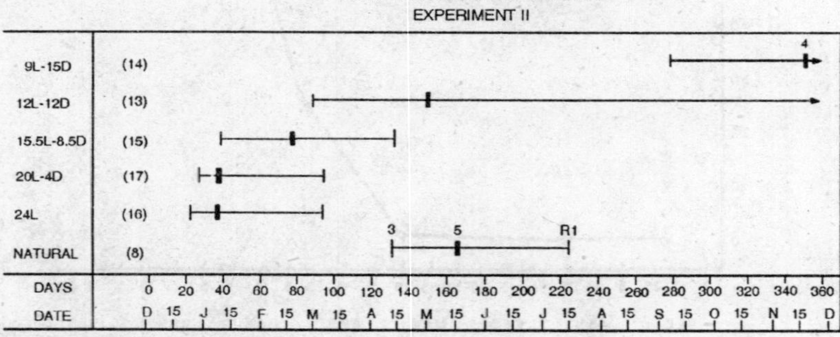

Fig. 7.4. Testis cycle in relation to photoperiod. Numbers at left of horizontal bars indicate the number of birds at the beginning of the experiment.

migration in birds wintering in the tropics and in the southern hemisphere. Even though the days are constant on the equator and are decreasing in the southern hemisphere after December 21, they could be well above the length of the daily effective photoperiod for the birds wintering there, judging from the studies of north temperate species.

Day length and the Maintenance of the Migratory and Reproductive Responses

This first series of experiments demonstrated primarily a relation between photoperiod in a 24-hour cycle and the initiation of the fat and gonadal responses. Does the daily photoperiod have a role in maintenance (or duration) of these responses once begun? If so, then the daily photoperiod would be significant in the regulation of the entire annual cycle, and not just in its initiation. Experiments were designed specifically to answer this question. Birds caught during the spring migration in April and May were subjected to constant day lengths of 9, 12, and 20 hours (photoperiods in a 24-hour cycle), and to natural day lengths. At the beginning of the experiment the gonads were partly developed, and the birds showed also heavy fat deposits indicative of the migratory physiologic state.

In the 9-hour group the gonads regressed almost immediately, in contrast to the previous experiment in which a gonadal growth response was induced with a constant 9-hour photoperiod. Was the reduction in day length as such responsible for the regression in the 9-hour group? Perhaps not, for in the 12-hour group, which also experienced a reduction in day length (from natural conditions) of approximately $2^1/_2$ hours at the start of the experiment, gonadal growth continued.

In the natural and 20-hour groups the gonads regressed (with one exception) after a few months of activity, as occurs in nature. In the 12-hour group, activity of the testes was maintained for about nine months; in the previous experiment (II), no gonadal regression was observed. Fat deposition continued in all of the groups, but the fat deposits disappeared first in the 9-hour group. The largest fat deposits were seen in the 12-hour group. The deposits disappeared (with one exception) in the natural and 20-hour groups before they molted, but most of the birds in the 12-hour group retained their fat deposits for many months and failed to molt.

The general similarity in the response of these groups and the natural group suggests that the gonadal cycle is operating at close to its maximal rate by late April and a marked increase in photoperiod

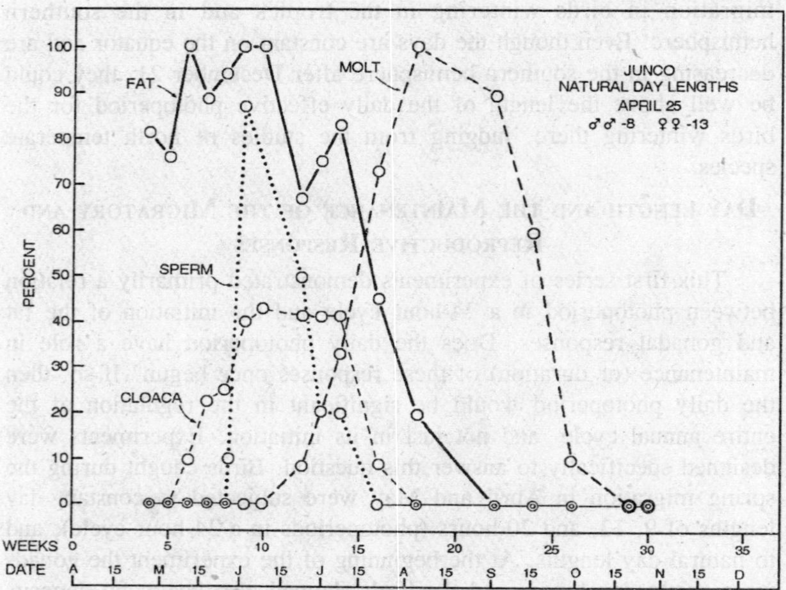

Fig. 7.5. Fat, reproductive, and molt responses in the slate-coloured junco under natural day lengths.

(from about 5 hours on April 25 and gradually decreasing to about $3^1/_2$ hours on June 21) had little if any effect. However, the birds treated beginning April 6 seem to show a slight acceleration of the gonadal and molt cycles.

The results of the first series of experiments showed that the daily photoperiod determined the time at which the response occurs. The results of the second series demonstrated that the daily photoperiod regulates also (a) the maintenance of the response, (b) the rate at which the response develops, and (c) the duration of the gonadal and fat cycles (or the time of gonadal regression).

Role of Light and Darkness in the Progressive phase of the Migratory and Reproductive Responses

At about the time the second series of experiments was concluded. Kirkpatrick and Leopold (1952, 1953) and Jenner and Engels (1952) demonstrated the importance of the dark period in the gonadal response to the daily cycle of light and darkness. Further experiments were undertaken by us to learn more specifically the role of the dark period.

If the effective stimulus is the total quantity of light which a bird receives daily, then birds receiving either 20 hours of light per day in

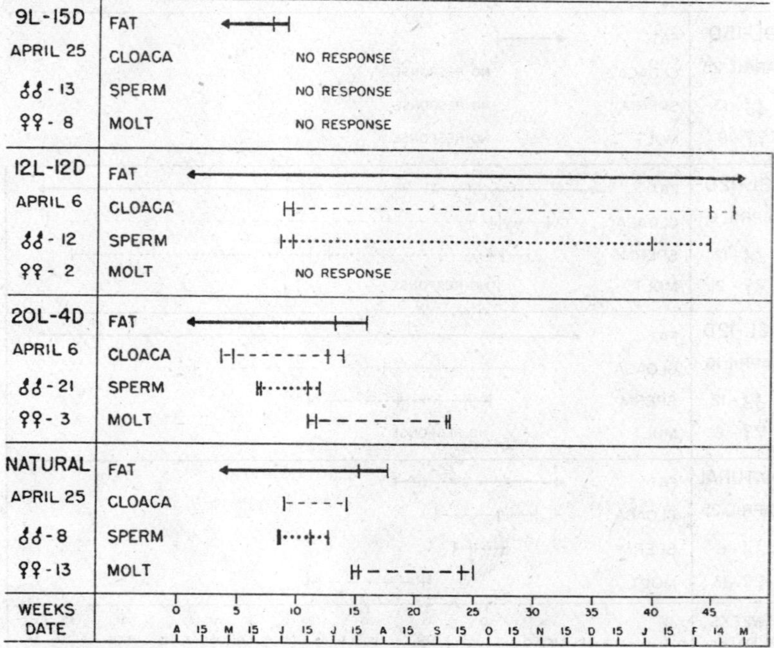

Fig. 7.6. Fat, reproductive, and molt responses in relation to phosoperiod.

one dose or four equal doses of 5 hours should respond equally. In the first experiment the birds were subjected to 5-hour photoperiods followed by 1-hour dark periods (*5L— 1D*). The schedules of Kirkpatrick and Leopold, *9L—7D—1L—7D*, and Jenner and Engels 8.25*L*—7*D*—1.75*L*—7*D* both involve interruption of a 14-hour night by a brief light period. In our test of this situation we employed a 8*L*—7.25*D*—1.5*L*—7.25*D* schedule resembling the preceding "interrupted night" experiments and an *8L—8D* schedule. Though the daily ration of light was 9.5 hours in one case and alternately 8 and 16 hours in the second, both of these schedules were highly stimulating. Thus, breaking the long night period precludes its inhibitory regulatory influence. An alternative proposal has been put forth by Farner, Mewaldt, and Irving (1953a, b). They suggested that the daily dark period *per se* has no positive function; rather, there is a persistent carry-over period which follows the end of each photoperiod, and the effective part of a photoperiodic schedule is the photoperiod plus the carry-over period. They envisioned the duration of the carry-over period to be a function of the duration and

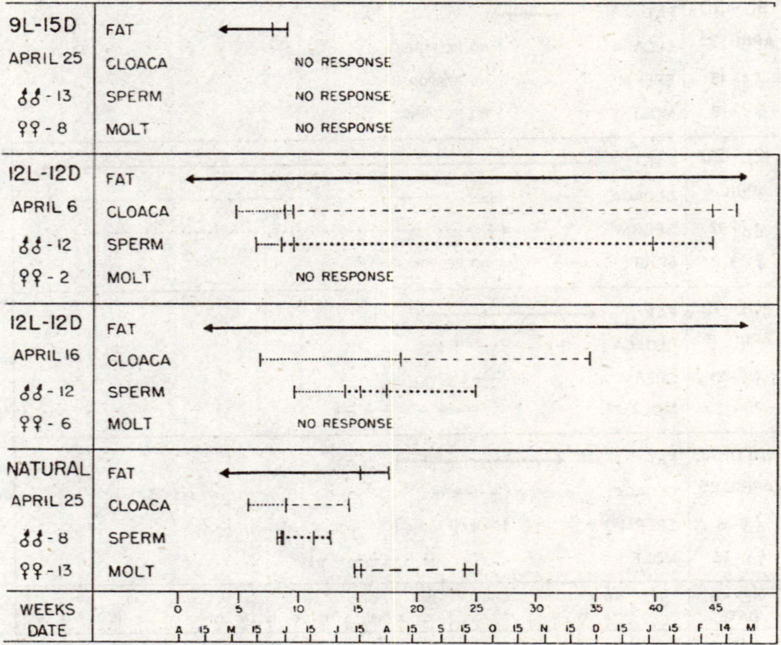

Fig. 7.7. Comparison of fat, reproductive, and molt responses in birds treated with 12L-12D, beginning April 6 and April 16.

nature of the preceding photoperiod. Their interpretation of the rate of testicular response as a function of the summated daily gonadotropic effects was in agreement with the summation hypothesis.

Experiments designed to test this thesis with *Z. albicollis* in our laboratory have been, in general, confirmatory. The schedule 1*L*—2*D*, which, in sum, equals the nonstimulatory 8*L*—16*D* simple light ratio, was strongly stimulating. This appears to rule out the mere ratio of light to dark as a controlling factor. However, it does not favour either of the two alternative interpretations that can be drawn from it: the carry-over hypothesis of Farner et al., or the interruption of inhibitory long nights as suggested by Kirkpatrick and Leopold, and others.

As a further test of the intrinsic properties of long nights we tried a (1*L*—0.25*D*) × 7—1*L*—14.25*D* schedule. This, in 24 hours, provided as much light as the stimulatory 1*L*—2*D* schedule, but it proved nonstimulatory. Neither did this schedule maintain breeding condition in birds already in the reproductive phase. Unfortunately,

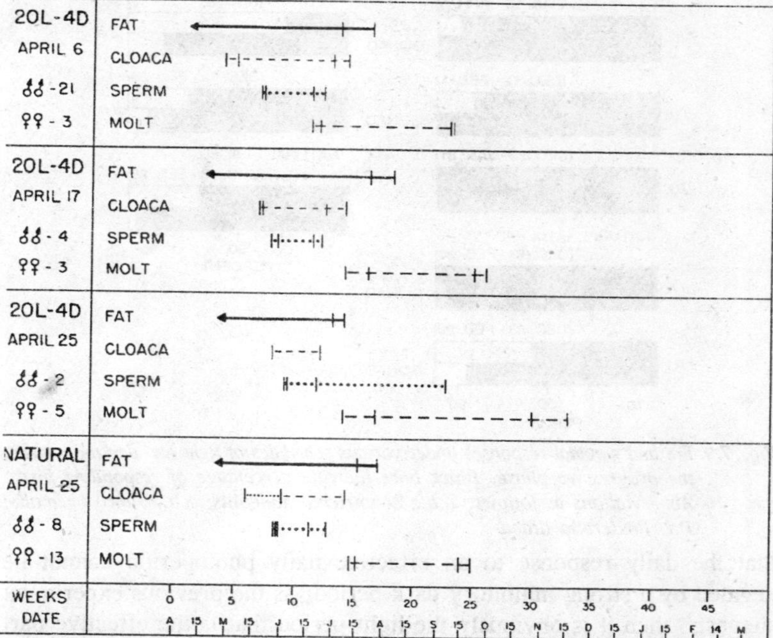

Fig. 7.8. Comparison of fat, reproductive, and molt responses in birds treated with 20L-4D beginning April 6, 17, and 25.

although this experiment appears to favour the long night inhibition hypothesis, it still does not rule against the "carry-over" idea, since quarter-hour interruptions of the 8*L* might not be considered long enough to allow for effective carry-over of light stimulus.

Further, we tested the following complex schedule (1*L*—2*D*) × 7— 1*L*—16*D*, an alternation of the 1*L*—2*D* stimulatory program with long nights. It proved stimulatory, and may, perhaps, be interpreted as denying the inhibitory function of long nights. However, this 38-hour cycle may not be sufficiently comparable to a 24-hour day to permit extrapolation. However, the effective light period, despite interruptions with dark periods, was 22 hours long and the dark period was 16 hours long. The proportion of light to darkness in this 38-hour cycle would have favoured a response. A similar view may be taken of the 12*L*—16*D* schedule which we found gonad-stimulating.

Experiments are now in progress employing longer dark periods, for example, 12*L*— 20*D*; 16*L*— 22*D*, 16*L*— 32*D*. If such studies show

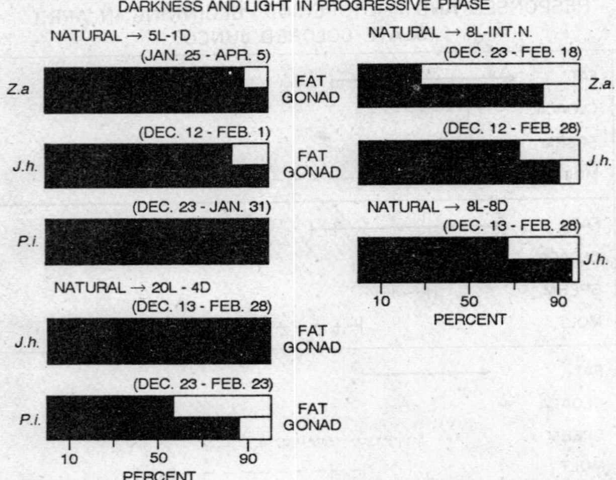

Fig. 7.9. Fat and gonadal responses under various schedules of light and darkness during the progressive phase. Black bars indicate percentage of responding birds. Abbreviations as follows: Z.a., Zonotrichia albicollis; J.h., Junco hyemalis; P.i. Passerella iliaca.

that the daily response to an effective daily photoperiod cannot be negated by a strong inhibitory dark period, as the previous experiment suggests, then it is obviously the light period that is the effective part of the photoperiodic cycle. The supposed inhibitory role of the dark period would then be merely a matter of semantics. A long night would be inhibitory only because it did not permit a longer or effective duration of light (photoperiod plus carry-over period).

Two additional schedules of light and darkness have been used: 16*L*—16*D* and 8*L*— 16*D*— 16*L*— 8*D*. The former was designed to test the inhibitory nature of a 16-hour dark period in a 32-hour cycle, the latter to test whether an effective daily response could be stored when administered on alternate days. Both schedules were effective.

It is still impossible at this point to distinguish precisely between the roles of light and darkness in the regulation of the gonadal and fat responses in the progressive phase. However, some things seem clear: (1) The total amount of darkness in a 24-hour cycle (when administered in small doses) is not the equivalent of the same amount of darkness given in a single dose. (2) Short cycles within a 24-hour period with the same proportion of light and darkness are equivalent to 24-hour cycles with regard to stimulation (5*L*—1*D* equals 20*L*—4*D*), but they are not equivalent with regard to failure to stimulate or "inhibition"

(1*L*— 2*D* and 8*L*—16*D*). (3) With different effective schedules of light and darkness within a 24-hour cycle, there is a difference in rate and duration of response. (4) Whatever the roles of light and darkness, the daily schedule of light and darkness in nature is the critical external factor. Effective daily schedules probably result in physiological "increments of response" which are eventually manifested, or at least become more readily observable and measurable, as gonadal growth, spermatogenesis, fat deposition, and increase in body weight.

Day length and the Refractory Period (Preparatory Phase)

After a period of seasonal breeding activity, the gonads regress spontaneously, sometime in July and August for most north temperate species. During this period, long days or increasing days cannot induce gonadal activity, and, hence, it has been called the refractory period. The natural termination of this period varies with the species, but it occurs usually in October or November. The refractory state can also be produced in the laboratory. Burger (1949) has reviewed the status of the problem of regulation of reproductive cycles and aptly stated that "attention has been focused too narrowly on the progressive phase of the reproductive cycle."

Juncos in nature experience a reduction in day length during the summer and fall, and it was plausible that decreasing day lengths, or short days, or long nights, could regulate the duration of the refractory period. Burger (1947) and Bissonnette showed that treatment with long days in the starling eventually induced a refractory period which could only be dissipated by treatment with short days. Miller (1948, 1951, 1954b) found that treatment with long days, begun in the fall in the migratory golden-crowned sparrow (*Zonotrichia atricapilla*), prevented the occurrence of gonadal growth at the normal time in the spring. This has been demonstrated by us for the junco and white throated sparrow. Birds refractory after long daily treatment with 20 hours of light were made responsive again to 20*L*—4*D* by six weeks of 9- or 12-hour light days. They remained refractory if kept in natural (longer) illumination during these six weeks. In related experiments, birds exposed to 20*L*—4*D*, beginning in the natural refractory period and extending for as long as six months, failed to respond, even in the normal spring breeding season.

These experiments indicated that it should be possible to induce responses several times within one year, even though juncos normally show only one period of gonadal activity and two periods of fat deposition

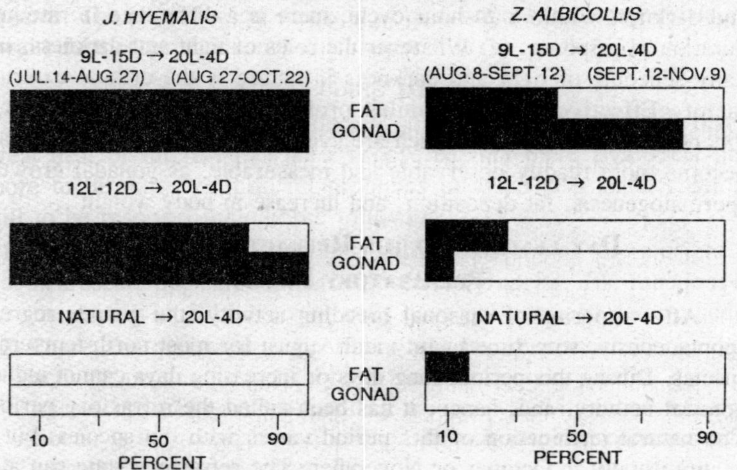

Fig. 7.10. Photoperiod in relation to completion of the preparatory phase.

annually. This was attempted by exposing birds to alternating periods of short days (*9L—15D*) and long days (*20L—4D*). Five periods of gonadal activity, five periods of fat deposition, and two molts occurred within the 369 days. Gonadal growth, fat deposition, and increase in body weight were correlated with long days. Gonadal regression, loss of fat deposits, and decrease in body weight were correlated with short days. These correlations became less distinct as the experiment progressed. Prior to this study, Rowan (1929), Miyazaki (1934), Damste (1947), and Burger (1947) had induced more than one period of gonadal activity in a year in different species of birds.

The results of this experiment confirmed the conclusion that short days in the fall regulate a reaction which enables the bird to respond to subsequent photoperiodic treatment. Without this "preparatory period" the bird does not respond. Although the bird is "refractory" to long days in the fall, it is undergoing a reaction which is regulated by day length (short days) and which is necessary for a subsequent response. Hence, it seems more appropriate to call it the preparatory period or phase. This period is then followed by the progressive phase of the cycle.

Role of Light and Darkness in the Regulation of the Refractory Period

Having established the requirement for short days during the preparatory period, the next problem was the determination of the effective part of the short days. Was it the short photoperiods or the long dark periods? A variety of schedules was used during the natural refractory period which was followed by subsequent treatment with long days (20*L*—4*D*) beginning in December or January.

The results demonstrated that it was not a short period of light *per se*, nor the total amount of separate doses of light and darkness in a given day, nor the proportion of light to darkness in a short cycle that was effective. Rather, it was a single period of darkness, 12 hours long, or longer, in a 24-hour cycle that was the effective part of a short day. A 16-hour dark period was generally more effective than a 12-hour period, which seems to be near the threshold for a minimum effective duration of the dark period.

The next question was whether a 16-hour dark period *per se* was the effective stimulus, or whether a long single dose of light was inhibitory. To answer this, birds were exposed to schedules of 16*L*—16*D* and (1*L*—2*D*) × 7—1*L*—16*D* during the preparatory period and then treated with long days (20*L*—4*D*). Both schedules were ineffective and corresponded to long days. That they were actually long days was established by the response of birds to those same schedules during the winter, as described earlier. The results indicated that a 16-hour dark period *per se* is not the effective stimulus. There is some relation

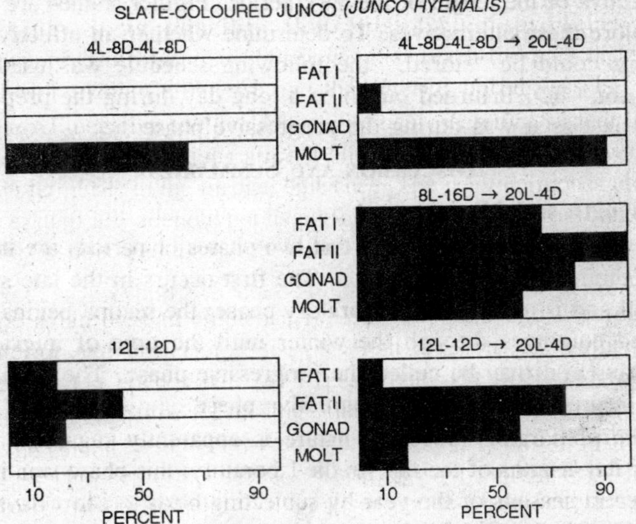

Fig. 7.11. Effectiveness of various schedules of light and darkness in inducing completion of the preparatory phase in the slate-coloured junco.

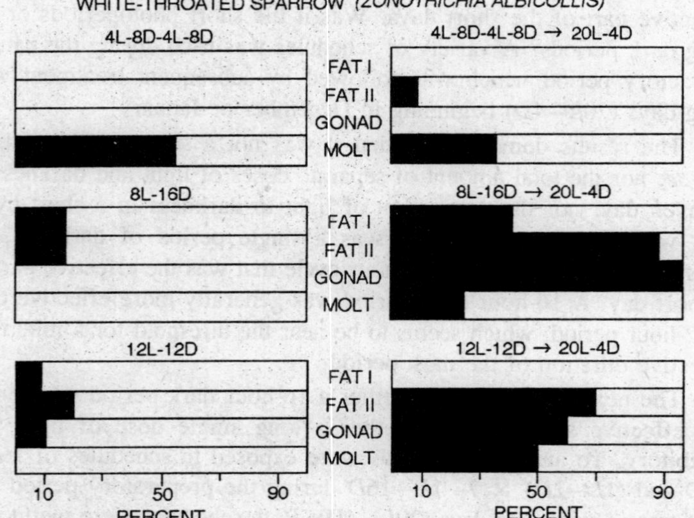

Fig. 7.12. Effectiveness of various schedules of light and darkness in inducing completion of the preparatory phase in the white-throated sparrow.

apparently between the duration of the light period and the duration of the dark period, or there is a single dose of light which is inhibitory irrespective of the length of the dark period. Further studies are needed to explore these alternatives. To determine whether an effective daily stimulus could be "stored," the following schedule was used: *8L—16D—16L—8D*. It turned out to be a long day during the preparatory period just as it was during the progressive phase.

Discussion and Summaries

Day Length and Migration

Available data demonstrate that two phases or periods are involved in the timing of spring migration. The first occurs in the late summer and fall and is termed the preparatory phase; the second begins in late fall and continues through the winter until the time of migration in the spring and may be called the progressive phase. The preparatory phase is prerequisite to the progressive phase.

The preparatory phase in nature is apparently regulated by the shorter day lengths of the fall. In the laboratory, this phase was induced at different seasons of the year by subjecting birds to short days (*9L—15D*). Twelve hours of light per day appears to be near the threshold for the maximum day length which can act like a short day.

Long days (16*L*—8*D*) arrest the preparatory phase and the appearance of the subsequent progressive phase. Therefore, it is the short days of fall that may be assumed to regulate the migratory behaviour and breeding cycle which appear six to seven months later. Relevant data are few, but suggest that there may be some relation between the length of the dark period, the duration of treatment, and the rate at which the preparatory phase proceeds.

After the preparatory phase has been completed, the progressive phase begins. In nature, it probably begins automatically in late November and December when the days are short, but *the rate at which it proceeds* is governed by day length. The spring premigratory physiological state was induced by means of long days about 80 days ahead of its natural occurrence, in a majority of the birds. With a short day (9*L*—15*D*) the same state appeared in about 160 days, which was 40 days later than its occurrence in nature. The near-maximum rate of response occurred with a day length as short as 15.5 hours; the minimum rate occurred with a day length of 9 hours, but shorter day lengths were not tried. Nine hours is about the shortest day to which juncos are exposed in nature.

It is important to emphasize here that the progressive phase, though slowed, developed *when there was no increase in day length*. About 60% of the birds released in June after being held on 9-hour days since December undertook a migration. Increasing day lengths, whether or not they are gradual, are not necessary to induce spring migration. The role of day length, once the birds are ready to respond, is the regulation of the rate at which the response proceeds.

Once in the premigratory physiological state, the length of time birds remain in this state is a function of day length. Long days dissipate the state more quickly than short days, and under short days (12*L*—12*D* and 9*L*—15*D*) the birds may not lose their fat. A change from long days to short days will also dissipate the state rapidly.

In nature, the spring premigratory physiological state ends when the birds arrive on their breeding grounds, or shortly afterward. After the breeding season, the gonads regress, the birds molt, and subsequently, there is a physiological change, which precedes the onset of fall migration. Nothing is known about the factors which regulate this state. When the fall migration gets underway in September and October the day lengths have reached a length which is effective for the beginning of the preparatory phase of the next spring migration. And, thus, a new cycle begins.

Role of Light and Darkness in the Migratory Cycle

The available data suggest that in the preparatory phase, the effective part of the short days is the dark period. It seems likely that there is a dark-dependent response which requires a daily *uninterrupted* dark period of at least 12 hours duration. If the dark-dependent reaction goes on during short dark periods at all, it cannot summate to give an effective daily response, or a response after the lapse of many days. A long light period can negate the positive effect of a long dark period; hence, the interpretation of the role of darkness is provisional.

In the progressive phase, both the light (stimulatory) and dark (inhibitory) periods have been suggested by different groups of workers as the controlling part of the photoperiodic cycle. Our data do not permit a clear-cut choice between the two possibilities. This may be because both light and darkness play important roles in the daily photoperiodic schedule.

An interesting difference between the "light and dark" reactions of the preparatory and the progressive phases is that in the preparatory phase, where the dark period seems to be the critical factor, the effects of shorter dark periods in a 24-hour cycle do not summate to give an effective daily stimulus. In the progressive phase, where the light appears to be the most critical factor, the effects of short light periods do summate to give an effective daily stimulus when there is no inhibitory duration of darkness in a 24-hour cycle. In the preparatory phase a long light period *per day* appears to be inhibitory, whereas in the progressive phase, a long dark period *per day* appears to be inhibitory. In both phases, the data show clearly that the 24-hour cycle is highly significant. It is possible that an inherent 24-hour rhythm, or other innate rhythms in the bird, are related to the migratory and reproductive responses to photoperiodic schedules and this must be explored in future studies.

Day Length and Reproduction

The regulation of the reproductive cycle in the male also occurs in a preparatory phase and progressive phase. The role of day length in each of these phases is similar to that in migration, except for the much slower rate of gonadal response to short days.

The duration of reproductive activity and, hence, the time of regression is also regulated by day length. The shortest durations were produced by long days, the longest by 12-hour days. The long days of summer probably initiate and maintain refractoriness. Some data suggest that regression may be an innate phenomenon, whereas refractoriness may be a feature induced by the long days of summer in temperate latitudes.

The role of light and darkness in the reproductive cycle is generally similar to that in the migratory cycle.

Day Length and the Timing of Spring Migration in Equatorial and Transequatorial Migrants

The relatively constant day lengths of the equatorial region can no longer be regarded *a priori* as nonregulatory. Constant photoperiods of 12 hours have been shown experimentally to be effective, and the duration of the photoperiod, moreover, regulates the rate of response. Hence, birds wintering in the equatorial region could be responding to the relatively constant day lengths of 12 hours. The birds that cross the equator and winter in the subtropics or temperate regions of the southern hemisphere are exposed to gradually increasing day lengths after they arrive in late October or November and to gradually decreasing day lengths after December 21, which reach a length of about 12 hours on March 21. Therefore, they are exposed during their entire stay on the wintering grounds to long days which, although they increase gradually to a maximum and then decrease gradually to 12 hours, would remain at an effective photoperiodic level, judging from our experimental work with juncos.

However, the main problem in equatorial and transequatorial migrants is, perhaps, not the effect of the day lengths on the wintering grounds, but the relation between day length and the preparatory phase and the initiation of the progressive phase. Since juncos could complete the preparatory phase on 12-hour days in late summer, it seems likely that equatorial and transequatorial migrants would be exposed to a sufficient number of 12-13-hour days during the fall migration to complete their preparatory phase. It is interesting to note that in nature, juncos winter where they will experience dark periods which last longer than 12 hours; spontaneous initiation and rapid development of the progressive phase are thus assured. The regulation of the preparatory phase and of the initiation of the progressive phase is the critical problem in the regulation of spring migration in equatorial and transequatorial migrants.

Day Length and Breeding Cycles in the Tropics

It has not been possible to extrapolate directly from experimental studies of the temperate zone gonadal cycle to avian breeding cycles in the tropics. Studies of breeding cycles at all latitudes led Baker (1938) to the conclusion that "the main proximate causes of the breeding seasons of birds in nature are thought to be temperature and the length of day in the boreal and temperate zones, and rain and/or intensity of insolation near the equator." Recently, a re-examination of the problem led to the same general conclusion that day length and temperature in the higher latitudes and humidity and rainfall in the tropics are correlated with the breeding seasons. The significance of these climatic factors is believed to lie in their effect on food supply. Breeding seasons are regarded as an adaptation and apparently are timed so that the young can be reared when food supply is at a maximum. If the breeding season is, indeed, adapted to environmental conditions operating toward its close, other factors must be postulated for the initiation of the cycle. Although day length and temperature are acceptable for higher latitudes, in the tropics both "are too nearly constant to offer a possible explanation". Our experimental findings, however, suggest that day length should not be ruled out as a regulatory factor in the tropics simply because it is relatively constant. The data from two series of experiments and other observations show that the gonadal and molt cycles of some tropical and equatorial birds, African whydahs and weavers (*Steganura*, *Euplectes*, and *Vidua*), can be altered by changes in day length.

A number of other correlations point to a relationship between day length and reproduction in the tropics: (1) In many tropical species which occur on both sides of the equator the breeding periods in the northern and southern populations are correlated with the seasons and, hence, occur at opposite times of the year. Frequently this is associated with the wet or dry season, but often it is not. (2) In many groups of birds, and even in a species with wide distribution, clutch size tends to be smaller in the tropics than in the temperate latitudes. In the domestic fowl egg laying (which is not homologous to clutch size) is greatly influenced by latitude. (3) The maximum size of active testes in tropical species is only 8 to 67 times the size of minimum inactive testes, whereas in temperate species it ranges from 267 to 2096 times the minimum size.

Miller (1954a) has shown that in the uniform environment of the Magdalena Basin in Colombia (3° 12' north latitude) eight out of ten species studied showed acyclic and uncoordinated breeding; two species showed coordinated and cyclic breeding. All of the species showed "evidence of possessing the same innate mechanism basic to cyclic breeding as north temperate species. This consists of need for rest, or assumption of a refractory state, and a basic tendency to progressive recrudescence." Miller (1954a) regards tropical day length, or light stimuli, as sufficient to surpass threshold needs in the innate mechanism, and as a regulator of the duration of refractoriness, rate of recrudescence, and duration of breeding condition.

Marshall and Disney (1956) have tested the response of an equatorial bird, *Quelea quelea*, to increased photoperiods, and found some response to long days.

Keast and Marshall (1954) have made extensive studies of reproduction and breeding seasons in relation to drought and rainfall and found that the gonads may remain inactive for a succession of seasons during a prolonged drought. Desert species can respond quickly to rainfall, or its effects, and nesting may begin within a few days of heavy precipitation, irrespective of day length and light increment. Marshall and Disney (1957) on the basis of an experimental study of the induction of the breeding season in a xerophilous species concluded that an internal rhythm of reproduction exists which is modifiable by external conditions, including possibly rainfall and social stimulation. In their opinion it is unlikely that the breeding seasons of truly equatorial vertebrates are controlled by photoperiodicity, and many equatorial and other species have evolved a reproductive response to rainfall or its effects. The data enabling one to evaluate conclusions with respect to the effect of rainfall on reproduction unfortunately are still conspicuously few.

In summary, the relation between day length and the gonadal cycle in tropical species is not known, but the point that must be emphasized is that day length cannot be ruled out *a priori* as a fundamental regulator of migration and breeding cycles in the tropics simply because it is relatively constant. There is no reason to believe that all of the progressive and regressive aspects of reproduction will be regulated by day length in precisely the same manner in all species. Nor must all of these aspects be regulated by day length. Regulation of a few is all that is needed for day length to be a primary factor in the timing of breeding seasons. Other environmental factors, such as rainfall and diet, and psychic factors may prove to be important modifying factors with different degrees of effectiveness in the preparatory, progressive, and regressive, phases. Only day length has been shown so far to be a primary regulatory factor, and its relation to gametogenesis and reproductive rhythmicity in tropical species remains to be determined by extensive experimental studies.

Neuroendocrine Mechanisms in Relation to Regulation of Migration

From the excellent and ingenious experiments of Benoit and his collaborators primarily, the mechanism whereby light induces reproductive activity during the progressive phase in the duck is thought to be as follows. Receptors for light are in the retina or in the encephalic regions near the orbit, the hypothalamus and the rhinencephalon. Stimuli from receptors reach neurosecretory cells of the paraventricular and supraoptic nuclei in the hypothalamus, but the pathway from the receptors to the neurosecretory cells is not known. From the hypothalamus, Gomori-positive neurosecretory material moves down the axons to the median eminence and the capillaries of the hypothalamo-hypophyseal portal system. When the material reaches the anterior lobe of the pituitary, the pituitary secretes gonadotropins which induce development of the gonads. During dark periods, the gonadotropins are retained in the hypophysis.

With this kind of mechanism, the different rates of gonadal response to different photoperiods could be explained by different rates of synthesis and secretion of gonadotropins. But how can we explain rapid responses when only 8 hours of light in 1-hour doses are given per 24 hours and no response when the 8 hours of light are given in a schedule of 8*L*—16*D*? Farner, Mewaldt, and Irving (1953) have suggested that the response to light may involve a process which becomes active almost immediately and which continues to exert its effect after the lights go off. Therefore, the daily gonadotropic effect would be the summated gonadotropic effects of the photoperiods plus the periods which occur in the darkness and called the "carry-over' periods. With only one 8-hour period of light and one carry-over period there apparently is insufficient gonadotropic effect. With an additional light period during the night, a greater gonadotropic effect is produced. Another way of thinking of it is that interruption of the dark period (before the effect of an 8-hour light period falls below a threshold value) produces an effect as though the light period had not ended. Obviously, very little total light is needed each day to give an effective gonadotropic stimulus, but this light must be administered in such a way that the duration of a single dark period is not 14 hours or longer in the junco. The effective stimulus appears to be the light; darkness seems to be inhibitory only in that it prevents light from acting.

After activity has been maintained for some time, the gonads regress. This regression could be induced by a feedback mechanism involving the sex hormones, the pituitary, and the hypothalamus, but attempts to inhibit the "light response" of the gonads with sex hormones have not been completely successful. Farner (1959) reports little or no gonadal suppression with testosterone or progesterone. Kobayashi (1954), on the other hand, was able to inhibit the photoperiodic gonadal response in the white-eye (*Zosterops japonica*) with either testosterone or estradiol. More important, perhaps, in this connection are the experiments of Bailey (1950) and Lofts and Marshall (1956) with prolactin. Bailey found that prolactin inhibited the photoperiodic gonadal response in *Zonotrichia leucophrys pugetensis*, and Lofts and Marshall induced testicular involution in the house sparrow (*Passer domesticus*), the greenfinch (*Chloris chloris*), and the chaffinch (*Fringilla coelebs*). Hence, prolactin could be responsible for the cessation of gonadal activity at the end of the breeding season.

The cause of refractoriness is not known, but it seems to reside at the hypothalamo-hypophyseal, and not the gonadal level, since administration of gonadotropins to either intact or hypophysectomized birds during the refractory period induces gonadal development. With regard to the pituitary, it is known to contain gonadotropic hormones during the fall, so refractoriness may be at the hypothalamic level if it controls secretion of gonadotropins from the pituitary. Kobayashi (1957) defines the refractory period as the period during which the hypophysis is secreting enough thyrotropin to induce molt, but in our laboratory we find birds refractory many weeks after the molt has ceased.

The preparatory phase might conceivably be a period of synthesis of gonadotropins in the pituitary, the mechanism requiring long periods of darkness each day, or probably more correctly, being inhibited or arrested by stimulatory doses of light. When sufficient quantities of gonadotropins have been synthesized in the pituitary, then perhaps secretion under hypothalamic control may be initiated during the progressive phase. The long days of summer, or of experimental treatment, possibly inhibit synthesis of gonadotropins, and, hence, prevent indefinitely the occurrence of the preparatory phase. A short treatment on short days perhaps returns the pituitary to synthesis of gonadotropins which then can be secreted under the regulation of the hypothalamus and photoperiod. That the situation is probably not so simple, is indicated by Vaugien (1955b) who showed that in immature English sparrows there is a gradual increase in responsiveness during the fall.

The role of the hypothalamus as a possible regulatory center of the mechanisms involved in the production of the migratory physiological state has been under consideration for some years, but we are still a long way from understanding the mechanisms whereby it exerts this control. Nevertheless, it seems likely now that the hypothalamus is the primary target organ of the external factor, the daily durations of light and darkness, and that the interaction between the external factor and the hypothalamus results in the occurrence of spring migration and breeding periods at the proper time of the year.

INDEX